Home Contractor's Secrets – Revealed

Contractor Reveals How to Save $1000's on Your Home Improvement Projects

By Matthew Miglin

Published by:
Chapelwood Publishing
Newton, NJ 07860

ISBN 0-9778581-0-3

Printed in the United States of America

Bulk Orders Available: 866-316-3700
Email: info@homecontractorsecrets.com

DISCLAIMER AND/OR LEGAL NOTICES

Acknowledgments

It took over four years to move this project from a great idea into a complete book. Through years of sweat and pain, I have learned the principles of what works and how to make and save money. In this book I share those principles and much more to help you, the reader, in your future endeavors.

I want to first of all thank my God for the strength, the ability, the desire, and the opportunity to help others through my books, audios, and speaking engagements. Without him in my life I could never do this.

I especially want to thank my faithful and beautiful wife, Karen, for standing with me and supporting me through the many hours and days I was stuck in my office in front of a computer, putting all this together.

I want to recognize the meticulous efforts of Teresa Douglas for her ability to translate my words into print, while capturing my heart and themes for this book. Her months of typing, editing, and organizing my thoughts and research into a book that flows has saved me months of additional work.

I want to thank the friends and family who helped with the book reviews, ideas, contacts, and testimonies to help make this book and the marketing of it a reality.

My heart's desire is to empower others to prosper and to know the truth.

To All Contractors

After twenty years in the construction industry, I have seen my share of good and bad contractors. I am sure you have, too. I have found that good contractors like us have no trouble finding work or finding good-paying customers. Many of our customers will never read this book, and those who do, will probably not apply most of these concepts to their home-improvement projects because they trust those of us who are good, reliable contractors. They will find it easier to let you make your markup and give them the quality service they expect. I believe this book is for home owners who always want to save a dime somehow and for the avid do-it-yourself home owners who generally won't hire contractors like us.

Contractors, I hope you read my book, too. It has a few ideas and secrets you might not know about. But you don't have to feel left out. I've written a book especially for contractors who want to make over six figures in their business. You can find it at http://www.sixfigurecontractor.com.

May your business double, and may you become more profitable.

Matthew D. Miglin

Table of Contents

Introduction

Can You Really Save Thousands of Dollars Using This Book?

I made a big promise at the front of this book. I promised to share my contractor's secrets with you—information the average home owner just doesn't have, information my fellow contractors don't want you to have. I promised that this information would save you a significant amount of money.

Can these secrets work for you? Absolutely.

Saving money on your remodeling project isn't rocket science. You don't need a college education or a record-breaking IQ. What you need is specialized knowledge. Knowledge that I am going to give you. Knowledge that will put me in hot water with my fellow contractors.

Why am I giving you this knowledge?

I'm not just a contractor. I'm a home owner, too. And though I lead a very comfortable life now, I didn't always. I know how it feels to count every penny. I know what it's like to need something I can't afford. And aside from buying your house, remodeling your house may be the single biggest purchase you will have to make.

I've seen too many people throw away money on remodeling projects that they don't like. I've seen people drive away honest contractors because they didn't know the right way to hire one. I've seen them get burned by the scam artists.

These kinds of headaches hurt good contractors as much as honest home owners. You have to worry about wasting your money. We have to worry about wasting our time. You worry about the contractor taking your money and running. We worry that you won't pay us at all. Instead of a straight-forward business arrangement, we have to navigate through suspicion and mistrust.

It has to end. I've decided to end it. And so you have this book.

This book will show you:

- How to spot a dishonest contractor from a mile away
- How to find a good contractor at a reasonable price
- How to find the best quality, least expensive materials in your area, and
- How to consistently save 5-50% off your home's remodeling costs

This book is not fluff. I'm not going to tell you how to remodel your bathroom yourself. I'm not going to give you the same old, tired information you can get from every other "remodel-your-home-for-less" book and pretend it's something new.

This book is going to give you information you need to know and can't find anywhere else. I know. I've looked.

What you hold in your hands is a summary of the important cost-saving secrets contractors have at their disposal. I'm going to assume that you've read some of those general remodeling books that you can buy anywhere. When I do stop to explain something basic, I'll let you know so you can skip that section. I'm not here to waste your time.

I'm here to save you a whole lot of money.

Read the book. Learn my secrets. Follow advice. Then let me know how much money this book helped you save. If you discover any new ways to save money, let me know. You may see your story in the next edition of *Home Contractor Secrets—Revealed*. If I use your story, I'll give you a copy of the book and a $50 reward.

How's that for a deal?

So what are you waiting for? Turn the page and learn how to play the home-improvement game ... and win.

Warning

Home Contractor's Secrets--Revealed is going to show you how to save thousands of dollars on your next home-remodeling project. It will show you how to pick the best contractor for the job and how to avoid getting burned by suppliers and Uncle Sam.

If you are looking for a no-effort 'make money while you sleep' miracle cure, then don't bother reading past this page. This is a book for people who know that you have to put in some effort to come out ahead. For people who don't mind putting in the time and effort.

If you aren't prepared to work, *stop reading*. This book is not for you.

Everyone else, read on. Let's save some money.

If You Want to Save Real Money You Need a Plan

I know what you're thinking. "I have a plan. I am going to remodel my kitchen/bathroom/master bedroom and then sell my house for more money."

This is not a plan. It's an idea. Ideas are good starting points, but ideas don't save you money because they aren't detailed enough.

Think of your home improvement project as a war. "Whoa," you might be thinking, "Aren't you going a little too far? I'm making by bathroom bigger not trying to take over China."

No you aren't trying to take over a country. But you need to think of your project as a war anyway, and I'll tell you why. You are about to enter—even if temporarily—the wild world of construction. *Contractors*—those people who do remodel work for a living—think of their business as a war. It is a business with a lot of competition where only the strong survive.

If you want to win in the construction game, you have to adopt the mindset. Contractors are waging a battle for business survival. You are waging a battle against ballooning budget costs and shoddy work. If you don't have a plan, you could spend hundreds of thousands of dollars on improvements that don't raise the value of your home or don't improve the quality of your life.

How do you develop a plan, you ask?

It All Starts with the Right Questions

How are you going to pay for your remodel? Are you going to save money little by little and then pay for everything with cash? Or are you going to borrow against the equity of your home? Many people borrow against their home. But ask yourself: Is this the right move for *me*?

What is your remodeling budget? Figure this out *before* you begin. It will dictate what you can do.

What is the purpose of your remodel? Are you primarily doing it to raise your property value so you can sell at a higher price? Or are you improving your home so you can stay in it for a while? This affects what kind of project you may choose to do.

If the improvement is primarily for your own use, then you can do whatever makes you happy. If that means turning your basement into a home office, so be it.

If you want to sell your home, you need to keep an eye on what will increase your home's sale-ability. It will be even more important to watch your remodeling costs. You don't want to spend more money than your house can ever be worth. In a later chapter, I'll list some of the best improvements you can make to maximize your return on investment.

When do you wish to begin? If you're in a hurry, you aren't going to save very much money. The secrets I reveal to you in the following chapters require time. *Your* time.
I'm not talking about five or ten minutes once a week. Some of my techniques will take a good chunk out of your day. Are you willing to take the time to save money?

When do you want to finish the remodel? If all you really want is to get your remodel done right away, you are going to have to pay extra. If you are willing to be a little flexible, you can save thousands of dollars. I'll show you how.

Are you going to do the remodel yourself, have someone else do it, or a combination of both? Do you need an architect to design plans for you? Are you skilled enough to build your project yourself? Do you have the *time* to do it yourself? If you work full-time, you will have to spend your free time working on your project, and the job will take longer.

Are you willing to dedicate your weekends to your project? Will you and your family mind living in a construction zone for a while? Talk to them, and pay attention to the answers you get. Your aim is to improve your house, not break up your marriage.

Will you need permits? If you ask the city in which you live, the answer will be yes. Then again, when was the last time Uncle Sam *didn't* want more of your money? Now, I can't list every possible remodel that needs a permit, but here is a general guideline: You need a permit if you are going to do anything with the wiring of your home or if you move walls or windows. Basically, you need a permit if you are doing anything beyond a cosmetic change, including turning your garage into another bedroom, doing plumbing, any structural work, adding a deck, or foundation work. Some towns require permits for other things—do your research.

Will the city find out that you turned your garage into another bedroom? Maybe, maybe not. But if you elect to build that room without a permit, you will not be allowed to count it as a room when you sell your house.

You heard me. If you turn your garage into a third bedroom without first getting a permit, you have to sell your home as a two-bedroom. That means your return on investment will be low because you can't charge extra for the third room and your property will be less appealing to a potential buyer.

When you are interviewing contractors, ask them if you need permits for the kind of work you want to have done. We will cover how to be sure a contractor isn't lying to you in another chapter.

What is the purpose of your remodel?
Answer this question *before* you start construction.

Will your project be a high-end or a budget project? Again, what is the purpose of your remodel? If you want to sell your home for a higher price, then look for the best "bang for the buck." If you plan on staying in your home for more than five years, you'll want something that will age well and stand up to daily use and abuse.

Now That You Have the Questions, Answer Them!

Yes this is a lot to think about. Maybe you are feeling a little overwhelmed. But don't skip this step! Remember, construction is war. A general who doesn't prepare ahead of time loses lives, or, in your case, money. You have a lot of money riding on the decisions you make. The more time you spend preparing for your remodeling project, the fewer expensive mistakes you will make.

Involve the Troops

As you start answering these questions, involve your spouse or partner. The time to find out that you can't agree on what needs to be done is *before* you invest any time or money. Your family is going to live in a construction zone for a while. Who is going to be in charge of cleaning up the dust? How will you keep the kids away from all those shiny, sharp things? Will you have to live somewhere else for a while? Most people do. Where are you going to live? Will the remodeling interrupt the children's schooling? Can you organize the remodeling project into easy-to-live-with stages? Will your spouse or partner do any of the remodel work? How much?

If you can't agree on everything, be prepared to compromise. Your family has to live there, too. If you don't get agreement right away, your family can make a stressful situation even more miserable. This way if your wife, or example, complains about being tired of living in a construction zone or at a hotel, you can remind her that the improvements you agreed on are worth the added stress.

And remember that compromise is a two-way street. Let's say your wife wants a Jacuzzi bathtub. You say it's too expensive. To get the tub, she agrees to be in charge of construction clean-up during the remodel. Everyone wins. She gets her tub, and you don't have to bust out the vacuum cleaner after a long day at work. And because she agreed to be the cleaner, you can't get in trouble for not doing it. And that, my friend, is worth a few hundred bucks.

The more buy-in you get at the beginning, the easier things will go for you. I can't guarantee that you won't fight about the remodel. I can promise that you'll have fewer fights because you both agreed on the work that needs to be done.

Involve the whole family in the remodeling project.
Compromise where necessary.

Put it in Writing

Once you sit down and think about these questions, put the answers down in writing. This will give you a clearer idea of what you want and will help you discover any holes in your plan.

Writing things down is critical when there are other people living in the home that is going to be remodeled—especially if you and your spouse or partner have decided to compromise on any aspect of the remodel. Your remodel project will not be done in a day. One or both of you will forget what you agreed to do. Writing down everyone's responsibilities will cut down on the misunderstandings.

Give Yourself Enough Time to Do the Job Right

If you want to get your remodel job done right the first time, you need time. Time to plan; time to find good people; time to find high-quality, fairly-priced supplies; time to determine what needs to be done; and time to organize yourself and your project. I know I've said this before, but I'll say it again: You need to invest some of your time if you want to save a lot of your money.

Some of you might be thinking that your time is too valuable to waste. You might be thinking that if you spend some extra money to get your project done faster, you'll save the thing you can't earn— extra time.

This myth really annoys me and I'll tell you why. Let's say that you decide to spend an extra $10,000 to have someone else complete your remodel job in a few weeks. You figure that you'll just work a little longer at your job and use all that time you aren't working on your remodeling project on other, more relaxing activities.

I don't have to tell you that the $10,000 has to come from somewhere. Let's say that you make $40 an hour at your job. To find out how many hours you need to work to earn that money, divide $10,000 by $40 an hour: $10,000 ÷ $40 = 250 extra hours of working. That's an extra thirty-one eight hour days you will have to work to earn $10,000.

Or is it? It's not as if Uncle Sam lets you keep all the money you earn, right? Every paycheck, some of your money is taken out for taxes, Medicare, social security, and whatever else you might have deducted. Now, I'm about to generalize here to make a point—your withholdings might be different—but, in general, let's say that for every $100 you make, you get to keep around $75 (75% of what you made).

That means, if you work 250 hours, you aren't going to get $10,000. You are going to get $7,500. Therefore, you need to work **335 hours** to get $13,400, which, after taxes, will leave you $10,000.

How did I get that number?

$10,000	Income
x .75	What you actually get to keep
$ 7,500	Left after taxes

$13,400	What you actually need to make
x .75	What you get to keep
$10,000	Left after taxes

This means that you would have to work even more hours to earn that $13,400 so you can afford to spend $10,000. Instead of working thirty-one extra eight hour days, you will have to work forty-two extra days to make that money. That's almost an extra month and a half.

Let me state it this way: You are going to have to spend more time at work to earn that $10,000 you spent to save time. Does that make any sense? It shouldn't.

Now let's say that you read this book and answer all of the questions and follow all of the tips. You shave $10,000 off the price of the remodeling project. But now you know that you are saving more than just $10,000. You are also saving yourself from having to pay all of that tax money you would have had to give to Uncle Sam if you just worked longer hours to earn approximately $13,400 to end up with $10,000. And you are also saving a month and a half of your life. My quote for this example is, "A $1.00 saved is a $1.34 earned."

This means that saving your money is more profitable and takes less time than working harder to earn that money. Remember that when you start to question if all this planning is worth it.

Are you ready to save a lot of time and money?

A Quick Review of the Remodeling Process for the Home Improvement Novice

In this chapter I am going to assume that you know nothing. If you have remodeled your home before, feel free to skip this chapter. For those of you with a dream and no idea how to make it happen, read on.

To keep things simple, I am going to assume that you want to remodel your home for your own use and not specifically to increase the value of your home. I am also assuming that you either have the money for the job or have applied for a loan.

Let's say you want to modernize your master bathroom and replace the carpet in your master bedroom with wood flooring. Now, since you don't have any construction experience, you need to bring in an expert.

The expert in this case is a *general contractor*. A general contractor is someone who has an overall knowledge of how a house is put together. More is involved, of course, but for our purposes, we'll keep it simple.

The general contractor can't do everything that needs to be done on a house, but what he can't do, he can hire other people to do. If your job needs an architect, the general contractor will probably know one. If you need a plumber, the general contractor will know a few that he has used before.

Besides locating skilled labor, a general contractor's primary job is to make sure your project is completed properly and on time. He will plot out, step-by-step, how each piece of your job will come together. He will tell the subcontractors when to show up and what to do. He will keep them on a timeline and get all of the necessary permits.

When the city inspectors come to make sure your house is being built safely, he will be there to answer questions and make changes.

If construction is war, then the general contractor is the five-star general. You are the commander-in-chief. Your job is to find a reputable contractor.

Three things should happen before you start interviewing general contractors:

1. Figure out exactly what you want him to bid on.
2. Read articles and books that talk about your type of remodel to acquaint yourself with the process and the terminology.
3. Draw up sketches and make a list of the materials you will need for your remodel and do some comparison shopping.

The general contractor plans each phase of your project.
He tells the subcontractors when to come
and what to build.

Three things should happen before the contractor starts working on your house:

1. You will sign a contract (more about this in later chapters).
2. You will give the general contractor a deposit.
3. The general contractor will apply for the necessary permits and hire the people necessary for the job.

The people the general contractor hires are called subcontractors. You pay the general contractor. The general contractor pays the subcontractors. If everything goes smoothly, you'll be living in sawdust for a little while and then have the bedroom and bathroom of your dreams.

In theory, you can know absolutely nothing, and the general contractor will handle everything for you. I don't recommend this. Theory and real life don't often resemble each other.

A *Little Knowledge Goes a Long Way*

Your general contractor could be the most honest person to walk the Earth. I'm going to teach you to pick honest, fairly priced contractors to work on your remodel project, but you shouldn't blindly trust your general contractor to make all of your decisions for you. That would be giving away too much of your power.

Even honest contractors make mistakes. If you don't educate yourself, you won't be able to tell whether someone messed up. Neither of you will catch the mistake. The contractor could be very sorry, but you are the one who will have to live with it—or, at least, live with the delay while the contractor fixes it.

You don't have to be an expert. All you need is enough information to ask the right questions.

If all goes well, you will be living in sawdust for a while then have the bedroom or bathroom of your dreams.

Three Ways to Educate Yourself before You Begin Remodeling

1. **Read home improvement books or magazines.** Libraries carry do-it-yourself home-improvement books that you can read for free. Home Depot has quite a selection that you can buy. You can go online to sites like www.contractor. com or www.servicemagic.com and look at their free articles. Familiarize yourself with what is involved in your project.

2. **Go to your state's official Web site and look up information on permits and contractor licensing.** This information will tell you what permits your future contractor is going to have to get for your project. You can also check to see what qualifications a general contractor has to have to work legally in your state.

3. **Find general contractors and ask them questions.** Do you know anyone who is a general contractor or is in construction? Start picking their brains right now. Buy them lunch and ask them about what goes on during a remodeling job and what they think you have to know before you get involved in your project. You will have a better idea of what to expect.

Wrap Up

Educating yourself will not take you very long, and will give you something to start with as you take advantage of the secrets in the rest of this book. While we are on the subject of educating yourself, let's answer a very important question...

What Kind of Remodeling Project Should You Do?

In the last section you knew exactly what you wanted to remodel. But back in the first chapter, I had you answer the question, "What is the purpose of your remodel?" As I said there, this answer affects the kind of remodel you should do.

If You're Remodeling Your Home So You Can Stay in It Longer

This is easy. If you are remodeling your home so you can stay longer, you have more flexibility. You don't have to worry about return on investment from a real estate perspective. My only advice is that you should figure out what will give you the most "bang for the buck," especially if you are on a limited budget.

That doesn't mean that putting down orange tiles in the kitchen is a good idea, even if you love the color orange. You may think you're going to live and die in the house you're in now. But things happen. Lives change. You may need to sell your house and go somewhere else. If that happens, you'll want to get as much money for your home as you possibly can.

Your decisions should be balanced between two things: On what will raise your home's value and on what will raise your quality of life.

Marble countertops are nice. Will your life be miserable if you don't have them in your kitchen? Perhaps your money would be better spent fixing your cabinets. Are you planning on having children? Perhaps you need to convert the basement into a bedroom or add a second bathroom.

If You're Remodeling to Increase the Resale Value of Your Home

Different people have different tastes. This should not be a news flash.

What you consider an attractive upgrade, someone else could see as a terrible waste of money. You don't want to spend thousands of dollars only to find out that your home's value hasn't gone up. This is a very important point, so I'll say it again. Remodeling your home doesn't automatically mean its value will go up.

Before you decide on your home-improvement project, find out what upgrades will give you the best value for your money. How do you do this?

Just from my own knowledge, I can tell you that kitchen and bathroom remodels tend to be good investments. In general, people like big kitchens and new bathrooms. A nation-wide statistical study released in May 2005 showed that a bathroom remodel increased the value of a house by 8.7%. Adding a bedroom only increased the value by 3.7%. Going strictly by the numbers, the bathroom is the better upgrade.

Square footage is also popular. On average, every 1,000 square feet home owners added increased the value of the home by one third. That is the average. Obviously, in big cities where space is tight, the square footage will fetch a higher price than in a rural area.

Sometimes you will have little choice about what you should remodel. If all the other houses in your neighborhood have two bathrooms and your house has only one, that lack will hurt your home's selling price. You might have to add a bathroom just to catch up with the neighborhood. As for bedrooms, a three-bedroom house always sells faster than a one- or two-bedroom house.

Pick up a copy of *Remodeling Magazine*'s annual "Cost vs. Value Report." This will tell you the average amount of money recouped from each home improvement. This is the Bible that realtors quote from.

Just to give you a preview, here are a few numbers listed in the 2005 report (Pg. 89):

Improvement Cost Recouped

Siding Replacement	108.6%
Bathroom remodel, midrange	102.2%
Attic Bedroom	93.5%

As you can see from this sample, some improvements are better than others. Siding replacement pays for itself and increases the value of your home. An attic bedroom almost pays for itself. Again, these numbers will vary by area.

Once you have looked at this publication—or before, if you have trouble getting it—you might want to bring a realtor to your home. He or she can look over your specific house and tell you what will help you raise the value of your home.

Don't Overlook the Cosmetic Changes

I won't spend much time on this. This book is going to deal with home improvement secrets. The following list covers information you should know. If you already know it, skip it. For the newbies in the audience, read the following:

1. **Give the outside of your house a power wash.** Basically, a painter or some other professional with a pressure washer comes by and blasts all of the dirt off the house.

2. **Paint.** Inside and out. The outside of your house should be a neutral color. Look through popular home magazines to get ideas for the inside.

3. **Landscape.** A little grass goes a long way. Plant some flowers and bushes because a good first impression will sell your house.

4. **Get your home professionally cleaned.** Your house could be the hottest thing since sliced bread, but no one will notice if your house is filthy. Even if you keep up with the housework, pay someone to give the house a thorough cleaning. Make sure your house smells clean as well.

Is everyone caught up on the no-brainer improvements? Good.

Remember Whose Taste Is Important

If you are remodeling your house knowing that you are going to live there forever, then you can build to suit yourself. If you want to put down green shag carpet and paint your walls orange, then go ahead.

Most of us, though, can't guarantee that we are going to stay in a place forever, and if you have to sell, you want to make the most money possible. This means that your house has to look good to potential buyers. This statement should be obvious. You want buyers—several buyers, just in case a deal falls through—to fall in love with your house.

Do you want the cold truth? Here it is: Nobody cares how much you paid for your house. Nobody cares that you raised your children in that house or threw wonderful parties. Your house holds sentimental value only for you, not for potential buyers. Unless your house looks good to other people, it isn't going to sell.

Honestly, if you want your house to sell, it has to look like all those pictures you see in Pottery Barn, or other home magazines. People want something fashionable. They want something modern. Even if they say they want something unique, nine times out of ten they want what they see in a catalogue. Give them that look, and you will have a very popular home.

Wrap Up

If you want to remodel to increase the value of your home, you need to plan carefully. Some projects will increase the value of your home. Others might be necessary because of your neighborhood. Some projects are a complete waste of time.

Are you planning on staying in your home for more than five years? You have a little more flexibility. But you should still consider how your improvements will affect the value of your home—unless you know for sure that you'll be living and dying in the home you are in. In that case, you can do whatever you want. Let your heirs sort it out later.

Be Specific

The more specific you are, the more money you will save. Period.

Once you figure out what kind of remodel you want to do—whether it is to help you stay in your place longer or to sell for a higher value—it is time to design your remodel. Do this before you interview contractors. The more specific your project is, the more accurate the bids will be. The fewer details and the more general your description, the more costly and inconsistent your bids will be.

Many resources exist that can help you design your remodel:

>**Stock plans.** These are pre-designed house plans. You can get them out of a "plan book," buy them on a CD-ROM, or find them on the Internet. Already drawn plans can cost anywhere from $350-$800, depending on the size and features in the home. Smaller houses have cheaper plans. A couple of places to try could be www.homefinder.com or www.abbisoft.com.

>**Builder plans.** Some builders have plans of the kinds of houses they usually build. Since builders are in the business to build houses—not sell plans—the costs are minimal. Often they are folded right into the cost of building your home.

>The disadvantage to these two options is that you can't make changes too easily. With stock plans, you are on your own. You have to hire someone to help you change what's there. Builders won't be too interested in giving you a discount to change their usual plans. They will make the changes for you, but that can cost $500-$1,000 and up for the time involved.

>Also, using a builder's plans cuts down on the number of specific bids you can get. The builder has no reason to give you those plans so you can get other bids. That means you won't be able to compare apples to apples—which is the

prime way you can save money right from the beginning.

The advantage in these two choices is that someone who knew what they were doing designed the plans. Hiring a draftsperson to draw out what you want may be cheaper, but there will be no one to tell you that the design is awkward or won't look great.

Architect. An architect can custom design your home, but you're going to pay for it. The industry average is 5 – 12% of the cost of your home. So if you are building a $200,000 home, your full-service architect may cost around $24,000.

One way to cut down on the expense but still have the architect's expertise is to buy a stock plan and use the architect to modify it, or you can develop your own plan and have the architect review it. Either option will cut down on how much the architect has to do, which will lower the price.

The amount of money an architect will charge you to do a plan review varies by area. Get a ballpark price before you decide to go this route.

Home Designer. A home designer is less expensive than an architect, but there are fewer services. The cost is, on average, around 2% to 8% of the cost of your home. If you are designing a $200,000 home, the home designer could cost around $6,000.

This is an excellent option when you want to custom design your home but lack the budget for an architect. Be aware, though, that some states will want you to get these plans sealed by an architect.

Draftsperson. A draftsperson generally doesn't offer much in the way of design services. That doesn't mean they lack the skill. Though not all draftspersons turn into architects, an architect generally has to serve some time as a draftsperson before gaining the title of architect. In most states, a draftsperson needs to have an architect review and sign off on the plans for them to be legal. It is a good idea to have an architect review the plans anyway. Hiring a draftsperson can cost anywhere from $500-$2,500.

Retired Contractor. A retired contractor is a great resource to have at your disposal. He may or may not be able to design your project for you, but he can tell you when things aren't going to work. He can also help you pick the most cost-effective materials and supplies, and point out other good contractors that can build your project. No matter whom you chose to draw your design, I recommend you use a retired contractor's experience if at all possible.

Lumber/Home Improvement Stores. Stores that sell construction materials to the public sometimes have design services. Home Depot and Lowe's home centers, for instance, have kitchen-design service centers where you can plan the design you have in mind with one of their "experts." The services at these places is generally free or low-cost if you are buying your materials there.

Bear in mind that home-improvement stores make their money selling you products. Unless you ask, you won't know how much training the design "expert" has been given. Did the person go to school, or did their "training" consist of learning how to operate the computer software that does the design while selling you on the most expensive cabinets?

Yes, using a design service in a place like Home Depot is cheap. But if you are locked into using their materials, you could be missing out on saving more money. You have to pick the design service you can afford, but if you automatically reach for the lowest cost design service just because it's cheap, you could be setting yourself up for higher costs all down the line.

How can an expensive expert be cheaper than a free service? I am so glad you asked.

Take Advantage of Expert Knowledge

1. **An expert will avoid amateur design mistakes.** An expert will know, for example, how much clearance a toilet needs so someone can actually sit down to use it. He or she will know where to best place the closets and storage spaces for maximal use.

2. **An expert has intimate knowledge about the materials that go into your remodeling.** Do you want to put in a hardwood floor? Wood may just be wood to you, but an expert can tell you which kind is the best deal for your budget and remodel goals. Hint: The cheapest wood is not always the best. But then again, the most expensive product is not always the best.

3. **An expert can make your plans ultra-specific.** This is true whether you have the expert design your remodel from the ground up or have them work with stock plans. They can put down exactly the kinds of materials that you want your remodel made of. This means you will have more accurate, lower priced bids. And fewer cost overruns. Why?

Construction is war.
Your mission is to be as specific as possible.

How to Save Money Before Your First Bid

When a contractor bids on a project, he has to think about three things:

1) How much the materials are going to cost.
2) How long it will take to build your remodel and install your materials (because this affects how much he must pay his workers).
3) What his ultimate profit will be.

This is harder if he doesn't know exactly what materials he is using. That doesn't mean he can't bid on your project. It just means he builds in wider allowances. When we talk about allowances, we're talking about the amount of money the contractor puts into his bid for things that the home owner still needs to choose.

If you are remodeling your bathroom, for instance, you might not have settled on the type of tub you want. You just know it's going to be a whirlpool tub. The contractor has to be prepared to pay for the kind of tub you want and pay for the labor to install it, so he will give himself a fudge factor. Since the contractor is not there to work for free, he will probably plan for an expensive tub that is difficult to install. The prices of whirlpool tubs can be anywhere from $550 to $3,500 and up. Your contractor is not interested in losing money. Which price do you think he will plan for in his bid?

Let's say, though, that you were smart and had the architect or a retired contractor write down that you wanted a Kohler Devonshire Bath Whirlpool with Left Handed Drain in the color almond. Now the contractor can find out exactly how much the tub will cost. Because he will have the exact measurements and weight of the tub, he will also know exactly how difficult it will be to install it.

Your contractor also knows that he won't have to spend time ripping out work he may already have done on other parts of the bathroom to fit the tub you chose. He will, therefore, pass this savings on to you in the form of a lower bid. If you want the $750 whirlpool tub instead of the $2,000 one that the contractor would have allowed for, you just saved yourself $1,250.

Let me say that again. You just saved over a thousand dollars *on just one product.* Think of all the money you will save if you specify all of the materials you want used in your bathroom or other remodeling project. It adds up to a lot. And can be well worth the price of that expert knowledge.

Not only will you pay less money for your project, but you can save time. If the remodeler or builder knows exactly what you want, he has more time to order the materials and plan how everything should fit together. That means that if you want a specialty item, it will get to your house on time—without extra rush fees.

And the contractor won't have to hold up work while he is waiting for it to come—or tear out work because your product doesn't fit to work he already finished. To save even more time and money, remember that white and chrome appliances and fixtures are usually stock items. They are available the quickest and usually have the lowest price of all the colored products. White works with most colors.

Make Your Decision and Stick with It

The subheading pretty much says it all. If you spend all the time necessary to pick out exactly what it is you want in your remodel, don't throw all of that away by changing things during construction. I am going to make this point again and again, but that is because it's important. Once your blueprints are finalized and you begin construction, don't change anything.

Being wishy-washy will waste your time and money. Don't do it. A subcontractor might suggest changes to you while he is doing his job, but don't listen. The time to make changes is *before* you start construction.

Contractors have their opinions. You should encourage your

contractor to suggest changes he thinks might benefit your project while the project is still on paper. Talk the proposed changes over with the person who helped you with the plans. If everything still sounds good, then change the plans.

There is another reason why you should stay away from changes during construction. Your plans have to be approved by the Home Inspectors in your city. Once those plans are approved, they don't want you changing them. If you do change them, you will have to get them re-approved by the city. This costs money. And time.

Some of you might be thinking, "How will they know?" I'll tell you how they'll know. The Home Inspectors visit buildings that are under construction. Not just big commercial buildings—every building. They will check to see if you are building according to city code and if you are following your blueprints. If you get caught making changes, the Inspector can "fail" your remodel and make the contractor do it over again, or he can stop the work until you get the new plans approved.

I hope I don't have to explain what a hassle that is.

Spend time figuring out exactly what you want.
Once you make a decision, stick with it.

Wrap Up

Any way you look at it, it pays to be specific. When you specify which materials you want the contractors to use, you automatically lower your building costs. When you use an expert to specify those materials, you take advantage of a wealth of knowledge you do not have.

Once you make your plans, stick to your guns. Being wishy-washy will only hurt your budget. Use experts to help you develop a set of plans that you can live with.

Of course, this means you have to know where to get your experts. Both the people you use to design your plan and the people who build your home have to come from somewhere. How do you find these people, and how can you pick the honest ones? Turn the page and find out.

How to Find and Hire Good People

The last chapters gave you a general idea about how the remodel process works. We covered what improvements have the highest return on investment and how to lay the groundwork for a money-saving remodeling project. This chapter will talk about how to hire good contractors that will give you dependable, high-quality work for a reasonable price.

Hiring a contractor can be the most difficult part of your project. Your choice can literally make or break your home improvement project. A bad contractor will cost you money and time—and will build an inferior product.

A good contractor is worth his weight in gold. He will show you ways to get the most for your money, steer you away from beginner's mistakes, and, in general, make your life easier. So how do you find a good contractor? First off, there is something you need to know about all contractors:

Money is the Number One Motivator

Contracting is a for-profit business. These people aren't building bathrooms and kitchens out of the kindness of their hearts. They work to keep themselves in mortgage payments and groceries just like you, and, like you, a good contractor is looking for an honest day's pay for an honest day's work.

The mistake many home owners make is to hold onto their money too tightly. Yes, you want to make sure a contractor does all the work he signed up to do. No, you don't want to give someone money and watch him walk away with it.

But look at this from the contractor's point of view. He is going to do a lot of work for you, hoping that you are honest enough to pay him. Believe me when I tell you that every contractor, if he has been in business for any length of time, has been ripped off. Every one of us has worked hard, only to have a dishonest home owner decide not to pay—or been forced to do more work than we agreed on, just to get the money we're owed.

If the contractors you interview sound a little suspicious, it's because they've been burned before. It's nothing personal. They are just trying to protect themselves. How do you gain a good contractor's confidence? I'm glad you asked.

A good contractor is going to look for two things from you:

1. Can he count on you to pay him, and

2. Are you going to pay him on time?

If a contractor can answer yes to these two questions, he is going to feel comfortable taking the job.
I'm not suggesting that you give the contractor a huge deposit up front. In fact, DON'T give a contractor too much money up front. A typical deposit is 10-30% of the total project cost. Many states have a set maximum amount—usually around 30%. Check to see what the limit is for your state.

If a contractor asks for more than he is legally allowed, this should be a warning that you are dealing with a shady character. Run away. The only exception to the 30% rule is if the contractor asks for 50% of the materials cost up front. But then again, 50% of the materials cost will generally be less than 30% of the total cost, so there is no problem.

A good scenario is to give the contractor a third of the total cost up front, a third half-way through the project, and the last third at project completion.

An even better scenario is to give the contractor smaller payments more often. This is good for you for four reasons:

1. You will have to come up with less money all at once.

2. The contractor will be motivated to do an excellent job in a short amount of time. This is because the contractor knows that he will get paid after every phase of the job he completes.

3. Each payment period gives you the opportunity to check the quality of the work and stop any problems while they are still small. In the worst case scenario, you'll be able to fire a contractor before he does too much damage. This is actually a pretty good "contractor quality test." A contractor that does shoddy work is not going to be happy that you want to pay him as he finishes each phase of the job. A good contractor will be thrilled because he knows his work is good and will pass your inspection.

4. The contractor will be more willing to give you a price break if he will get paid more often. Remember, a contractor's biggest worry is that he's going to work and not get paid. He will work with you if he knows you plan on paying him throughout the project.

Let me state that last point another way. You will never get a contractor to lower his price if he isn't sure you're going to pay him and give him his money on time. You will have a hard time hiring a contractor at all.

And while we're on the subject, let's talk about the one sentence you should never EVER say to a contractor …

A contractor will feel comfortable working with you
if you answer his two important questions:

Will you pay him? Will you pay him on time?

"I'm Getting Other Bids"

This is the kiss of death. If you tell an in-demand contractor that you are getting other bids, he's going to write you off. You will never hear from him again.

Now, if you've done any reading at all on how to remodel your house, you're probably scratching your head right now. A lot of those books tell you that you should tell the contractor that you are getting other bids. But they're wrong, and I'll tell you why.

See, when you say "I'm getting other bids," you are trying to tell the contractor that you will know it if he inflates his prices. What the contractor *hears* is, "I want to make you work like a dog and pay you next to nothing, *if* I pay you at all."

A good contractor has a lot of job offers. He no longer *needs* to work like a dog for bad pay to establish his reputation. You have just become more trouble that you're worth.

You have to understand that making a bid is time-consuming work. A contractor doesn't just make something up and give you a price. He has to calculate how much your job is going to cost in materials, time, and labor and figure out his mark-up. Every job is different. There are no shortcuts. This means that a busy contractor is looking for reasons not to bid on a project because he doesn't want to waste his time.

If clients tell me over the phone that they are getting other bids, I find some way to cancel the home visit. If a client says this after I've gone to their home to discuss their project, I don't follow up. I don't want to spend my time developing a solid presentation and coming up with a fair price and then have you choose some fly-by-night company because their price is a lot lower. It's a waste of my time. It's as simple as that.

That doesn't mean that no one will offer to do your work for you.

Two kinds of contractors won't care that you're getting other bids:

1. Contractors new to their business, and
2. Scam artists, or contractors who do shoddy work.

Now, there is nothing wrong with someone who is new. We were all new at one time. If you want to take a chance on someone new, go for it. You may have a great experience, and you'll be giving someone deserving a chance to prove himself.

Just be prepared to watch the whole process carefully. If you are a complete novice to construction, I suggest passing on the new contractor. You don't know enough to catch any beginner's mistakes he might make.

There is another type of contractor who doesn't mind being the "low-price leader." Actually, they have to work for low prices. It's the only way they can get people to hire them without looking too hard at their past jobs. These are the contractors who don't deliver on what they promise. Who don't—or won't—take the time to build a quality product. Or they are the type who act as if their clients are morons.

Kind of funny, isn't it? There you are, trying to get yourself a good deal by telling a contractor that you are getting other bids, and it backfires. Lucky for you, I'm going to tell you the right way to get a good contractor for a reasonable price. But first, we need to be clear on something …

You Should Get Other Bids— Just Don't Tell the Contractors!

Doing anything else is stupid. You have a lot of money riding on your decisions.

And the fact is, you do want an excellent job done for as little as possible. A reasonable person knows that at some point, when you demand too low of a price, you are getting the shoddy work you paid for. A savvy person also knows that just because something is more expensive doesn't make it better.

So how do you tell the difference? How do you find that middle-range sweet spot where you get a great value for a reasonable price?

I'm so glad you asked.

Get Multiple Bids for
Every Contractor You Need to Hire

You heard me. And no cheating. When I say multiple bids, I'm talking about thirteen. Let's say you have to hire a plumber, an electrician, and a stone mason to do your project. You need to get thirteen bids for *each* of them. That's a total of thirty-nine bids for the job—unless you use a general contractor to do everything. Then you only need thirteen bids.

When I say bids, I am talking about your actually speaking to a live human being. Leaving a voice message on someone's answering machine doesn't count. You need to know how much money the contractor is going to charge you for your project. If you call during a contractor's busy season, you'll have to make more than thirteen phone calls because some of the contractors won't follow up with you. They'll be too busy.

What You Are Aiming For

You want the contractor to come to your house. Getting a bid over the phone doesn't cut it. Until the contractor sees what needs to be done, his bid is not going to be accurate. Well, how do you get a good contractor to come over and bid?

You present yourself as a serious customer. This is where all the work you did in the beginning will pay off. Let the contractor know that you have a set of detailed plans, have been approved for financing, and have a tentative start date. That will get his attention. That tells him that you are serious about your project and aren't just price shopping.

Settle on a time for him to come over. Once he is already there, let the contractor know that you would prefer a detailed breakdown of the costs to be included in your bid. He can do this if you have a detailed list of the materials that will be used on your project. This makes it easier for you to compare bids to see if all the contractors are giving you the same numbers for labor and materials.

Try to get separate prices for different sections of your project (i.e., bathroom, kitchen cabinets, bedrooms, flooring, etc.) but don't push for it. Some contractors, such as myself, will give detailed breakdowns of the project in my contract—but only one price. So don't push too hard for a price breakdown; otherwise, you will push away good contractors.

Once you have all this information, you can make an educated decision.

To Save Real Money, You Have to Put in Some Effort

Yes this is work, and it's going to take time. But remember, by taking the time to complete this step now, you are saving money, and an even larger chunk of your time. Remember our example from before? Saving $10,000 also saved a month and a half of your life.

Do the work, but don't announce what you're doing to the contractors involved. I'm not saying you should lie to them. Just don't announce the fact to the world.

An experienced contractor will often be savvy enough to ask you if you are getting other bids. Here is how you should answer them: "I am looking for the best contractor for the job. Money isn't as important to me as finding a high-quality person that I can feel comfortable working with. If you are that person, then you will get the job, even if you aren't the lowest price."

Practice your answer ahead of time so it sounds natural. When you answer this way, you can admit that you are looking without looking like a Scrooge.

Why am I putting you through so much work? When you get thirteen bids, two important things happen:

1. **You will see how much your job *should* cost.** The majority of your bids will cluster around a certain price range. Anyone wildly higher or lower can be safely eliminated.

2. **You'll be able to separate the good contractors from the bad, even with no prior experience.** Each contractor is going to come in and present their proposed project to you. You'll see how they handle themselves. What they include in their contracts. How they treat you. How respectful they are of your home.

That's why thirteen bids per contractor is the absolute minimum number of bids you should get. Yes, this is a lot of work, but in the process you will get an education from each contractor. Some of us are teachers. We'll give you information that will help you figure out which contractors do and don't know what they're doing.

As you get better with the process of remodeling, you can get away with six to ten bids per project. Don't feel bad about not giving the job to someone who did all of the estimating work. That is the nature of the business and contractors don't expect to get every job they bid on. Don't be afraid to pay for estimates from reputable contractors. Good contractors will often charge for estimates, so paying for an estimate is not a bad sign—unless you're tight on money.

Most contractors are quite happy to tell you what their competition is doing wrong. If one contractor tells you something you aren't sure of, ask the other contractors you have interviewed whether you've been fed a line. If the other guy has it wrong, we'll tell you, in detail. This information can save you a fortune because it can save you from hiring the wrong person.

Getting thirteen bids takes time and effort.
But the money you save makes it worthwhile.

Wrap Up

If you remember nothing else from this section, remember these two things:

1. Get thirteen bids for every contractor you need to use. No cheating.
2. You will only hire a good contractor if he can be sure of two things:
   a. That you will pay him, and
   b. You will pay him on time.

Ready to learn where you can find good, reliable contractors? Read on.

How to Find Good Contractors

Of course, none of the previous information does you any good if you don't know where to find contractors to interview. Here is a list of the places you should look, with the most important listed first.

Referrals/Recommendations

Do you have friends or family members who completed a remodel? These are the first people you should ask for a recommendation. Look over their remodel. Ask them questions about their contractor. Find out what they did and didn't like about him.

- Did the contractor always show up when he said he would?
- Did the contractor finish the remodel on time?
- Did the contractor stay within budget?
- Did the contractor treat them and their house with respect?
- Overall, how do they like the job he did?
- How was the quality of his work and materials?

You might think of other questions. The great thing about getting a recommendation from a friend or family member is that they usually give you an honest assessment of the person who worked for them although you probably already know if they like their remodel or not.

Even if your friend hates the contractor and the job he did, it is worth finding out why. Use what you learn to protect yourself.

Material Supply Yards

What is a material supply yard? It is where contractors buy their materials. These can range from lumber yards that sell fifty different types of wood in large quantities to Home Depot or Lowe's home centers that sells to the general public. Either way, the second best way to locate contractors is to go to where contractors get their supplies. Ask the manager for contractors that they would recommend.

The people at lumber yards and home-improvement stores will be more experienced than your friends and family, and they will know more contractors. Because of this, you might be wondering why I make them my number two recommendation instead of number one.

The reason is that your friends and relatives have first-hand experience with specific contractors working in their homes. This is information lumber yards do not have. They know these contractors as customers who buy supplies from them.

That doesn't mean they won't have an opinion on who is easy to work with and who isn't. It just means you will have to take the time to make sure the contractors they recommend are right for your project.

But you were going to have to do that anyway, right? So talk to your friends and family first. Once you've contacted their recommendations, move on to material supply yards.

Online

Another great resource is the world wide web. There are literally millions of Web sites dedicated to helping you find a contractor. While I don't specifically endorse any one site or guarantee you'll find the contractor of your dreams, here are a few you might want to look at:

www.contractors.com

This very informative site lets you look for a contractor based on the kind of job you want to get done. It also has a cost estimator, articles to expand your knowledge, a loan center, and a help-line. All of their contractors are pre-screened for licenses and insurance.

www.servicemagic.com

This site also screens contractors, and has a database of articles on home improvement. The design gallery is worth looking at if you are still in the idea phase.

www.reliableremodeler.com

Again, this site offers pre-screened contractors. Their articles are a little more helpful than servicemagic.com, especially the "how-to" series if you are thinking of doing some of the work yourself.

Personally, I would recommend using more than one service. Both Service Magic and Reliable Remodeler advertise that "up to four" contractors will bid on your project. That means, at best, you are going to have to find nine more.

Ask Subcontractors and Other Tradespeople

Do you know anyone in the construction trade? Have you had work done on your home before? It doesn't matter if the tradesperson or subcontractor you used doesn't do the kind of work you now need done. Contractors know each other. They work together on bigger projects. The tradesperson you know could give you an honest, unbiased view of how good other contractors are.

You can find great contractors online.
Don't overlook this resource.

Real Estate Agents

These are people who need to know—at least in a general sense—about remodeling. They are the people who most often tell you what you should do to increase the value of your home. Because of this, they may hear about different contractors via the home owners they serve.

Architects

Architects work directly with contractors. They can tell you how easy it is to work with different people and what kind of quality work they do. Often they have contractors that they use all the time. You will still need to ask questions and see references, but if the architect likes working with a particular contractor, that could be good news.

Trade Associations

In general, contractors who care about their product are members of trade associations. You can call these places or visit them online and get a recommendation for a contractor in your area.

You can look in your handy-dandy yellow pages under "Trade Associations," do an Internet search under the words "Trade Association," or go to such sites as www.nahb.org.

Your best bet of all would be to use all seven avenues we discussed: (1) get referrals from friends and family, (2) get recommendations from material supply yards, (3) use online services, (4) talk to subcontractors, (5) talk to real estate agents, (6) talk to architects, and (7) look up trade associations. Doing these seven things will give you a whole range of contractors.

Yellow Pages

I don't really recommend using the yellow pages in the telephone book because ads can be deceiving. Ad size has no relationship to the contractor's quality. It is a good resource if you need more local contractors for bidding, but you take more risks with these sources.

Here is a little fact you probably don't think you need to know: With the exception of painters or a plumbers, the majority of home owners don't use the yellow pages to find home contractors.

Why am I telling you this? For all I know, you are one of those people who go to the yellow pages for everything. But there are only four types of contractors (who aren't painters and plumbers) whom you will find in the yellow pages:

1. New contractors who haven't yet figured out that they just wasted their money.
2. Slightly more experienced contractors who have to wait until their subscription expires in a year before they can cancel it.
3. Contractors who are looking for clueless home owners.
4. Good contractors who are looking for credibility and/or exposure (the exception, not the rule).

Obviously, not all contractors who appear in the yellow pages are bad. However, it is hard to tell the good from the bad. An experienced contractor won't use the yellow pages because home owners who call are mostly price shoppers and time wasters. They are looking for rock-bottom prices. All and all, this isn't the best place to find winners.

Now that you know where (and where not) to find contractors, we'll cover the cheapest time of year to hire a contractor.

A Contractor's Least Expensive Time of Year

Can you guess when that time is? I hope you didn't say early spring. That tends to be the height of a contractor's season. The weather is warmer, and home owners have all that tax return money to spend on projects around the home. If you try to get bids in early spring, you will be paying top dollar—if you can get the contractors to call you back at all. Most contractors are swamped during this time and are looking for reasons not to make bids.

In winter, all of that changes. As the temperature drops, so does the demand for contractors.

This has always been funny to me. Home owners don't do remodeling in the winter, even when they aren't the ones working in the cold. The construction trades that get hit the hardest are deck builders, roofers, painters, and masons, but that doesn't mean every construction company closes for four months all across the country. We have to pay our bills, after all. Some work can still be done.

If you have an indoor project, or even a workable outdoor project, you can take advantage of this slow period. Some very good contractors—people who might otherwise be too expensive for your budget—get laid off during the winter. Or if they own their own company, they are looking for ways to pay their monthly overhead.

This means that the contractor will be willing to do a project for less money just so he can have money coming in. You get a better contractor than you could otherwise afford. He brings in money during a slow time. Everyone wins.

And here's an added secret for you: The slowest of the slow times is usually leading up to and through Christmas and New Year's. And for small jobs, the days leading up to and after holidays like Thanksgiving, Memorial Day, Labor Day, July 4th, as well as back-to-school time tend to be slow for contractors as well. This is because holidays distract home owners from considering doing work on their homes.

Most people don't want a construction zone in their house while they are rushing around during the holiday season. And by New Year's Day they are spent out. A smart contractor has a cushion to tide him through this lean time, but he won't pass up an opportunity to add a little more money to his bank account.

The only exception to this that I can think of is a contractor who does blown-in insulation. It's best to use him in the summer when no one else is thinking of staying warm. You can save as much as 50 percent doing this off-season.

And that is the key secret. Can you arrange your schedule to take advantage of super-slow times? It will take some long-rang planning. But you won't be sorry you did it.

Wrap Up

With a little searching, you will soon find many contractors to interview. To get the widest range of contractors, get referrals from friends and family, recommendations from material yards, and online Web sites.

If your project allows for it, schedule your remodel in the winter or small jobs around holidays when contractors are hungry for work. By taking advantage of these two secrets, you will find the best contractor for your project at a price that you can live with.

Or will you? How do you make sure the contractors aren't lying to you? How do you identify the scam artists?

For the answer, turn to the next chapter.

How to Eliminate 98% of the Scam Artists—Before You Hand Over Any Money

I hate to say it, but not all contractors are honest. There are people out there who will overcharge you and give you garbage. Others aren't dishonest, just incompetent—they couldn't build a birdhouse, let alone a bathroom, if you put a gun to their head.

Wouldn't it be great if all dishonest, incompetent contractors were terrible at selling themselves? That way, when you talked to them over the phone or had them present their project idea in your home, you could separate the good from the bad right away.

Yeah, and I'd like to win the lottery while we're dreaming. It is a sad fact, but …

A Contractor's Presentation Skills Don't Always Match His Building Skills

Let's throw in a scenario. Picture, if you will, two contractors:

Contractor number one, Joe Shmoe, comes to your home to talk about your project. He's all dusty and he stinks because he came to see you after a long day at work. He walks around your clean home leaving dust and the smell of his body odor everywhere.

He doesn't say anything and is looking somewhere else while you tell him about your remodel. When you are done he says, "I can do that," and pulls some pictures of his past jobs out of his wallet. Then he quotes you a price.

By this time you are about to pass out from his stench. You tell him you'll think on it and send him out the door. When he leaves, you open all the windows and doors and turn on the fan.

Contractor number two, Mike Moe, arrives an hour later. He is in a pair of crisp khaki pants and a polo shirt with his company's name on it. He doesn't stink.

Mike asks intelligent questions when you talk about your project. Afterward, he leads you through a glossy photo album of jobs. He answers your questions quickly and cheerfully.

Which Contractor Would You Want to Hire?

If you said stinky boy, there is something wrong with you. This is not a trick question. I'm not leading up to a big moral where I tell you that judging people by their looks is wrong.

We could come up with a million reasons why Joe Shmoe stinks when he comes over and why you should hire him anyway. The fact is, Joe is clueless. He is going to have to change his ways if he wants to do more than scrape by. But that's not the point. What is my point?

When you compare Joe and Moe, Moe looks like the better contractor. You're more likely to hire him *because he took a shower and brought over big pictures.* Your reasons have nothing to do with his ability to build something. In other words, you want to hire him because he presented himself better.

Scam artists know that if they present themselves well, honest people are more likely to hire them, even if they don't give you concrete information about their past jobs and references. Their presentations can be so slick you end up feeling like they are doing you a favor by agreeing to work for you.

That doesn't mean you should distrust a contractor who puts together a great presentation. Just don't assume that a great presentation means he will do a great job.

So how do you separate the professional contractors with great presentation skills from the scam artists with great presentation skills?

You Ask Questions!

If you contacted the contractor, this is your first step in separating the good from the slimy. Later we'll discuss contractors who call you.

When you interview the contractor, you need to keep control of the situation. That means you know exactly what to ask him when he walks in your door. Here are some of the questions you need to include in your interview:

1. How many similar jobs have you done recently?
2. Are you licensed and insured with liability and worker's compensation?
3. Do you have your own employees, or use subcontractors, or both?
4. How long have you been in business doing this type of work? (Seven years or more is a good, fifteen years is better).
5. How busy is your schedule? Will you be able to fit my project into your schedule? How soon?
6. Will you be on the job every day?
7. How many jobs do you schedule at the same time?
8. Who will be in charge and supervise the job every day?
9. What hours do you typically start and stop work?
10. Do you work weekends?
11. Will you cover and protect my personal items and furniture?
12. Can I look at similar jobs you have done recently?
13. Can you give me a list of referrals from other customers you've done work for in the past two years?
14. How reasonable were the contractor's prices for change orders and extras?

A slick presentation and friendly manner don't guarantee good work at a fair price. Be careful. Do your homework.

Listen carefully to the contractor's answers. After he leaves, call his references. Here are some questions you should ask:

1. How did you like this contractor?
2. Did the contractor meet his scheduled time frames?
3. Did the contractor show up on time each day?
4. Did the contractor meet his scheduled budget?
5. Was the job site cleaned up daily?
6. How did the contractor handle problems on the job?
7. Was the contractor good with details?
8. Did you feel comfortable using this contractor?
9. Were the employees and subcontractors comfortable to have around?
10. Would you hire this contractor again? Why or why not?
11. Was there anything missing from your house afterward?

Once you do all of this work, you will have a detailed picture of the contractor's work practices. Now, there may be situations where you don't bother to call the contractor's references. It may be that you can spot that the contractor is bad news. How do you do this? You look for the top eight warning signs that mean you have trouble on your hands.

If the contractor you are interviewing falls into any of these categories, cut the interview short as politely as possible and move on.

Warning Signs that You Might Be Dealing with a Crooked Contractor

1. **The contractor makes a cold-call to your house.** A "cold call" means he calls you out of the blue. He doesn't know anyone you know, and no one recommended him.

   If a contractor is making cold calls, it's because none of his other clients will recommend him or give him referrals. Do not work with this kind of contractor.

There are a few exceptions to this rule—some good contractors are aggressive in trying to grow their businesses, but they are rare. One way might be if your neighbors are remodeling their home—a worker might come by and give you a flyer. In this situation you can go see the remodel for yourself. You can ask your neighbor about the quality of the work. In this situation the worker isn't trying to sell you anything. If you like what you see, you can give the company a call.

2. **The contractor tries to pressure you into making an immediate decision**. An honest contractor knows you need time to think it over. Crooked contractors will tell you that their price is a discount that is good "only if you accept it by the end of the day." They are trying to stop you from getting any other bids. 90% of the time, their "discount price" is more expensive than everyone else's regular quote. Run, do not walk, away from these "deals."

3. **The contractor offers discounts or finder's fees for finding other customers**. Good contractors rely on referrals and word of mouth to get their business. Good contractors don't have to bribe you to make you work with them.

4. **The contractor wants you to get all the building permits.** This could be a sign that the contractor is not licensed and is trying to avoid getting caught. This is not always the case, but be aware.

The contractor's references can tell you how well the contractor deals with stressful situations.

5. **The contractor doesn't want to sign a contract.** He tells you that he's a man of his word and a handshake is enough. A handshake isn't enough. Verbal contracts are valid in some states. But how do you prove that the contractor went back on his word?

6. **The contractor doesn't have many references.** Anyone can find two references. For all you know, the contractor's friends are posing as clients. Ask for ten references. If he can't produce them, why not? He may be new to contracting. But then again, he might just be a lousy contractor. Don't relax if the contractor gives you ten references. CALL THEM. Most people don't bother calling references. Bad contractors know that. Call all the references and ask questions about the job they did. Was the client satisfied? Would they use the contractor again?

7. **The contractor says you have to pay for the entire job up front, or demands payment in cash.** Don't use this person. Just don't. A deposit of 10-30% is plenty. Pay by credit card or check so you have proof that you gave the contractor money.

8. **The contractor wants to handle the financing for you.** Never go through your contractor. Your contractor's job is to remodel your home, not act like a bank. People who used their contractors and signed paperwork in a rush find out too late about huge interest rates and all those points on the loan. It has happened before. Don't let it happen to you.

If remodeling is a war, then scam artists are the vultures that circle the battlefield looking for easy prey. A scam artist doesn't want to work. He wants money for nothing. But a scam artist can't fool you if you do your homework. Read this book, apply the secrets I'm going to give you, and you will see those vultures coming from a mile away.

Now that you know how to spot the thieves, here is how you spot the contractors you should hire.

Eight Signs That You Are Dealing with a Great Contractor

The contractor doesn't pressure you. A good contractor knows what he's worth. He also knows you need time to decide. A good contractor will give you that time because he is confident he is the best person for the job.

The contractor listens when you talk to him. Some contractors will listen to a few words, decide they know what you want, and then stop listening. A good contractor never assumes—he listens and asks questions.

He isn't desperate. A good contractor has a steady stream of work because word of his skills has gotten around.

He tries to educate you. A good contractor takes pride in his work. He will listen to what it is you want to do and make respectful suggestions about the project.

The contractor returns your phone calls promptly and answers your questions respectfully. In other words, he is a good communicator. Do not hire a contractor who makes you feel uncomfortable or stupid when you ask questions. A great contractor knows that this is part of the process.

The contractor has a clear set of guidelines in place. He will have a contract that is fair to both sides, and a clear procedure he follows for each project. He will be comfortable sharing this with you.

When you ask for references, the contractor gladly offers them. A great contractor has great references. He will want you to call them.

The contractor is safety conscious. Does the contractor have liability insurance? Is the coverage high enough to cover damage to your home and contents? I recommend $500,000 as a minimum that any contractor should have before the contractor is allowed to work on your home. If the contractor has employees, does he carry workers' compensation?

Think beyond simple insurance. A safety-conscious contractor operates his construction sites in a safe manner. He will have safety procedures that his workers must follow while on the job. Ask to see them.

If the contractor doesn't keep his workers safe—the people that bring in money for his business—why would he keep you or your home safe?

How good is his safety record?

A good contractor is open and honest about his business. He is a good communicator, doesn't pressure you, and encourages you to check up on him. A good contractor runs a smooth operation—but he has plans in place to deal with difficult situations that might come up. This contractor won't quote you a rock-bottom bid. But he will complete your project on time with a minimum of headaches. And that will make him a good value for his price.

Wrap Up

Use these checklists. They will help you to separate the good contractor from the scam artist. After that, you have to listen to your instincts. After interviewing thirteen or more contractors, you will be educated enough to make a good decision. Weed out the scam artists first. Then go with your gut. Pick the contractor who gives you piece of mind.

Now that you have the right contractor for your project, I have a very important secret to share with you …

How seriously does the contractor take safety?

Protect Yourself

A contract is your first and last defense in your home improvement war. It is the map that you refer to during the remodeling process. It is the armor that protects you in changing circumstances. And the weapon you use if you ever have to take your contractor to court.

How does a contract do all of this? I'm glad you asked.

The Contract Is Your Map

A good contract is detailed. It should lay out, step by step, what the contractor is going to do and when he is going to do it.

Be explicit. According to the Oxford English Dictionary, explicit means "clear and detailed, with no room for confusion or doubt."

You don't want confusion or doubt about your project. A lot of your money and the contractor's time are on the line. You and the contractor need to agree on what it is you are paying him to do. Write everything down. Assume that if it is not written in the contract, then it won't be done.

The contract should also describe *your* responsibilities. Are you buying the materials for the job? Does someone need to be home to let the contractor in? Your biggest responsibility is payment. If you listen to my advice, you are going to pay the contractor in small installments during the job. Your contract should state when the contractor can expect those installments.

Be specific. "Contractor shall receive first installment when 25% of the job is completed" is not specific. What do you mean by 25% of the job, anyway? "Contractor shall receive $500 once all carpeting is pulled up" is much easier to understand. The task the contractor has to complete is specific and so is the amount he will get.

Here are just a few of the things you need to include in your contract:

- A start and end date
- Who is responsible for cleanup and debris removal
- A detailed payment schedule
- A description of how change orders and extras are handled
- A description of who is responsible for obtaining permits, and who will pay for them
- Who is responsible for insuring the construction project if fire, theft, or loss occurs

This is not a complete list. The more detailed you make your contract, the better. If you would like to purchase a copy of the contract I use in my business, go to www.sixfigurecontractor.com/contract.

How a Detailed Contract Can Save You Thousands of Dollars

To make this simple, let's pretend that you are going to get your kitchen and bathroom remodeled and add an attached deck. Let's say you did your homework and called thirteen bidders and ended up contracting $49,500 for the whole job.

What if I told you that a specific contract could have taken a good $10,000 off of your price? How, do you ask?

The average home owner is not experienced when it comes to remodeling his or her home. The architect or general contractor who helps you design your remodel will try their best to be as clear as possible and spell everything out, but they need your specific input in order to do that.

Most people don't have the patience to sit down and spell out exactly the kind of trim and plumbing fixtures they want in their home, so these things are left for "later." In other words, the home owner chooses these materials when the contractor is already working on the house.

This means that those plans you paid for only cover around 60-85% of the actual job. As a result, your contract only covers 60-85% of the total job. This is a mistake. I'll tell you why. This isn't a repeat of the first chapter where we discussed how much you can save on the actual materials you buy. We are going to talk about how being specific saves you on labor costs as well.

But first let's have a more detailed overview of how a contractor comes up with his bid.

How a Contractor Comes Up with His Price

You call the contractor and tell him about the job you want done. He may come over and look at your home. Then he goes back to his office, takes the information you gave him, and figures out how long it will take him, in man-hours, to build your project. Then he multiplies that by the hourly rate of the people who will be working on the job—let's say, $35 an hour. If your project takes 400 man-hours to build, that comes up to $14,000. He will then add an overhead percentage, or markup, to this number.

The contractor will use this as the basis for his bid. He may have to add other numbers to this figure—he may be buying materials or other things—but the foundation of his price is based on his educated guess of how long it will take to build your project. When the contractor goes to build your project, he is very motivated to get it done within those 400 man-hours.

Got it so far? Okay. Now let's say you decide to add another set of cabinets into your new kitchen. The contractor is going to charge you an hourly rate since this is a new thing you've added to the project. Do you think he is going to charge you $35 an hour?

Make your contracts as detailed as possible.

No. Remember, the contractor has to add in a percentage for overhead to cover his costs, and the contractor will make that percentage higher just in case you change your mind again. Remember, he has to cover his costs AND make a profit. You'll probably end up paying him somewhere around $55 an hour.

Not only that, but the contractor is no longer in a hurry. He is charging you by the hour, not with a flat rate. If you had included this little project into the bid, you probably would have had it done in ten hours, starting at $35 an hour. Now you're getting those cabinets in twenty hours at $55 an hour. You could have had the cabinets for $350. Now you're paying $1,100.

Do you see the power that comes from being specific? I tell you now that most home owners pay 35% more for their projects than they have to because they aren't willing to be specific. But this doesn't have to be you.

In case you didn't get it the last few times I said it, build your project on paper BEFORE you get any bids from contractors. Walk through each room that is going to be affected by the remodel. Look at every wall and every detail. If you hired an architect, brainstorm with him or her until you know exactly how your project will look when completed. If you don't have an architect, find a retired contractor to go over the plans with you. You should know which materials you need, down to the hinges on the doors.

Don't feel overwhelmed. If you don't have access to a contractor's experience for free, it is well worth your money to buy a few hours of someone's time. We're talking about a $10,000 difference here. This is one of those rare occasions where spending a little money can save you a lot of money.

The Contract is Your Armor

Writing everything down protects you. If the contractor fails to complete a task, then he doesn't get paid. It is as simple as that.

Building in small checkpoints throughout the project gives you a chance to catch any problems when they are small and easily fixed.

It is also a good idea to have the contractor sign an unconditional lien release for the amount of each payment you make to him. This means that—as long as you pay the contractor in full—you are not responsible for paying off the suppliers and subcontractors. I know of horror stories where the home owner paid the contractor in full and then had to pay the subcontractors and suppliers when the contractor skipped town or defaulted. They can come after your house. Protect yourself.

So why would a contractor want to sign a contract? It sounds like the home owner gets all of the benefits. But that isn't so. A detailed contract protects the contractor as well. He knows what he has to do and what to expect from you. While he is promising to do a good job, you are promising to pay him a fair wage.

The ideal contract is one that protects both of you. Your contractor will (or should) have something drawn up once you decide to pick him. Review the contract. Does it protect both of you or just the contractor? Does it mention the payment schedule and map out when and how the two of you will communicate?

You have the right to negotiate the terms of the contract before you sign it. Make sure you understand what it is you are agreeing to. Don't be afraid to ask questions.

A contract should make your working relationship easier. It should try to spell out everything about the project. And give you ways to deal with the unexpected. A good contract gives you peace of mind. It allows everyone to concentrate on fulfilling their end of the bargain.

Be suspicious of any contractor who doesn't want to sign a contract. I know I mentioned this in the previous chapter, but it's important. A contractor who won't sign a contract is either a crook or an amateur. Either way, you don't want them working on your house.

The Contract is Your Weapon

Sometimes you have to go to court. If you use the lists on how to identify good and bad contractors, chances are that you'll never have to face this.

Let's say that you didn't take my advice. You only interviewed two contractors and chose the one who seemed like a good guy. You didn't call the contractor's references. You didn't spend much time interviewing him—the contractor is the brother-in-law of a friend and you made up your mind to use him before he even came to see you.

If you had done your homework, this contractor's previous clients could have told you that the guy is a flake. He doesn't show up when he says he will. Nothing gets completed on time. And there is always something wrong with his work.

Fortunately, you made sure that the contract detailed every step that the contractor was supposed to complete. You didn't avoid the shoddy work and frustration that comes with working with a flaky contractor, but at least you have something definite to charge him with when you go to court. You can say to the judge, "My contract said I was going to get ABC, but what I received was XYZ." This makes it easier to get a settlement in your favor.

Of course, you're going to take my advice on how to pick a good contractor and avoid all of this in the first place, right?

Use Your Contract Wisely

We've discussed the ways your contract acts like your map, your armor, and your weapon. Each of these functions has its place, but if you take enough time to use it as a map—spelling out all the project details—then you won't have to use it as weapon very often.

Your home improvement project is a business transaction, but it is also a partnership. If you use your contract as a weapon too often, you can kill the project. What do I mean by that?

Have you ever heard of late fees? Your video store isn't the only business that charges them. Contractors call them penalties. Some people put penalties in their contracts. If the contractor doesn't finish the project on time, he gets charged a late fee for every day he is late. The monies that a contractor has to deduct from his pay because of a penalty are called *liquid damages*.

This is standard practice in commercial work. A job is "commercial" if it involves building something where people aren't going to live. Building a Wal-Mart or Starbucks store is a commercial job. In this kind of work, millions of dollars are riding on a project finishing on time.

The company that oversees building these buildings is often called the project manager. The project manager pays a contractor much more than you can afford to pay. He is paying for speed.

If one contractor is late, it can derail the entire multi-million (or even *billion*) dollar project. If the contractor finishes early, he gets a large bonus for every day he shaved off his timeline. If a contractor finishes late, he gets fined.

Contractors who do commercial work have specific talents and temperaments. They know going into each project that they are taking a risk. The large amounts of money involved make it worthwhile to them.

Why Liquid Damages Are a Bad Idea for You

As you probably guessed, your project is not commercial. Your project comes under the heading of "residential." A project qualifies as residential work if it involves buildings where people live.

You don't have the kind of money that Starbucks has to play with;

therefore, you can't afford to pay the contractor enough money to make liquid damages worth the risk.

Besides, late penalties are not standard practice in residential work. A contractor may have chosen residential work specifically because he doesn't like working with a gun pointed at his wallet. Some people thrive in high-pressure situations. In my experience, contractors who work in residential don't like high pressure situations. Put yourself in their shoes for a moment.

Picture this: You walk into work tomorrow and the boss hands you an assignment. This is a type of assignment that you have done before. You know that you usually finish it in two days.

Today your boss tells you that you need to finish this assignment and have it on his desk in two days. For every hour that you are late, he is going to take $10 from your paycheck.

Ten dollars isn't much. And you know that you will probably finish the assignment on time. But didn't your blood pressure rise just thinking about what is at stake?

Losing money doesn't motivate a contractor. Insisting on late penalties will make you look too aggressive. Remember the two things a contractor is looking for in a client?

1. A client who will pay him, and
2. A client who will pay him on time.

When you put late penalties in a contract, you are trying to motivate the contractor to finish on time.

What the contractor sees is someone trying to find reasons not to pay him. That may not be what you're doing, but that is what the contractor sees.

I never work for people who want to put penalties or liquid damages into my contracts. It isn't because I'm afraid I won't be able to make a deadline. I can accurately estimate how long it takes to build something.

The fact is, when you are a good contractor, you can afford to pick and choose your clients. While you interview a good contractor to see if he is a good fit for your job, he is interviewing you. Will you be easy to work with? Are you going to try to micro-manage the project? Clients who want liquid damages fall under the category of "more trouble than they're worth."

Usually these are the clients who want to under pay you. They'll dangle the promise of a bonus under your nose if you finish early. The bonus might make the job profitable, but the contractor has to work twice as hard for it.

I want an honest day's work for an honest day's pay. I won't work like a dog, hoping I'll be fast enough to be profitable. I don't have to do that. There are other clients who will be easier to work for. You will attract the kinds of contractors who have to work under those conditions.

Choose a Contractor You Can Trust

Ultimately, you have to choose a contactor that you can trust to remodel your house. If you don't trust the contractor you are considering, adding late penalties is not the solution. Choose another contractor. Keep looking until you find one who is known for the qualities you find most important.

So, just for the record, I do not recommend late penalties. I think they will only backfire on you. If, for whatever reason, you are bent on adding late penalties to your contract, I'll show you a better way to phrase it.

Here is something that might work: If time is very important to you, you can try doing the opposite of late penalties. Try this when you and the contractor are negotiating price.

Let's say that your budget and his bid are pretty close, but he is still a little higher than you were hoping to pay. You want to spend $12,000 on your bathroom. You did your research and you know $12,000 is a fair price.

He wants $16,000. That is a little high for you, but there is something about this contractor that you really like. So offer him $12,000. Tell him that, once he finishes the project on time, you will give him a $4,000 bonus.

Now, he may or may not take your offer. But at least this way it sounds like he could earn more money. Late penalties phrased any other way focus the contractor on all he has to lose.

Wrap Up

Contracts are in that category of things that save you money before you have to spend it. A good contract acts like your map—it details where you should be at each phase of your project. Spend a good amount of time making sure your map is as specific as possible.

A good contract is your armor. It allows you to catch mistakes and misunderstandings while they are still easy to fix. Communication is the oil that keeps your armor smooth. Make sure you set down ways you and the contractor will communicate—and when. Make sure you communicate often.

A good contract is your weapon. If you spend enough time mapping out your project and protecting yourself against misunderstandings, you will not have to use it. But if something goes wrong, your contract will help you win the fight in court.

As a general rule of thumb, adding late penalties to your contracts backfire. If you don't trust the contractor, find someone else.

Do your homework when you search for a contractor.
Then go with your gut.

So far, the secrets I've shared with you have shown you how to pick your perfect project, an honest contractor, and how to protect yourself with a contract. These are all ways to save money before you even have to spend it. The next chapter can shave thousands of dollars off your next remodel project. Some of it is information you are not supposed to know.

Read on …

How to Save Money
on Your Up-Front Costs

Here's another fact for you to memorize: The contractor doesn't have to collect sales tax from you if your project qualifies as a "home improvement." If your project is a repair, then he has to charge you sales tax.

Often, the only thing that separates a taxable project and a non-taxable project is the knowledge of the people writing the contract. There is a certain wording you need to use. Make sure your contractor knows how to work this.

I can hear someone out there wondering if the contractor will help you avoid this unnecessary expense. Aren't they making less money if they don't collect sales tax from you?

The answer is no. The contractor has to collect sales tax for the government. It is an extra step for him, involving paperwork and sending money to someone else on a regular basis. And who likes to collect money for someone else? A contractor, therefore, will be happy to skip this step if it is unnecessary. Ask.

One Area Where You Will Always Pay Sales Tax— And How to Save Money There, Too

Even if your project qualifies as an improvement, you will pay sales tax on the materials used to build your project. You won't see that sales tax if the contractor buys all the supplies for the job. But you're paying for it just the same.

Remember, the contractor isn't building your project because he's a nice guy. He's doing it to make money. That means he is going to pass all the costs of the job on to you.

One Thing Contractors Don't Want You to Know

I'm probably not shocking anyone when I tell you that the contractor passes all of his costs on to you, then adds in some extra. This is called his *markup*. This gives him something called a *profit margin*. After the contractor pays all of his bills for the project, this is what he gets to take home.

You would be surprised at how small of a profit margin most contractors live on. Costs are high, and most of the money you pay goes straight to the bills. There is one exception, however. Can you guess what it is? Probably not, so I'll tell you.

Materials. When I talk about materials, I mean the actual *stuff* used to build your project. If I'm building you a deck, those materials would include such things as lumber, nails, and cement.

Since I'm a smart contractor, I can get these materials inexpensively. I know where to look. But I also make it a point to get thirteen bids on all of the materials I buy so I can be sure I'm getting a low price. You see? I don't ask you to do anything I am not willing to do.

Most home owners don't know how much materials should cost. Because of this, I can put a larger *markup* on top of the actual price of the materials than I can on anything else. This isn't as crooked as it sounds. That profit margin is a safety cushion—*your* safety cushion.

Let's say, for example, that my job is to paint the inside of your house and put on a nice trim. You pick the paint color and the trim. I go out and buy it, and I paint your house.

So far so good. Now what happens if I then put on the trim and run out of the material part of the way through because I underestimated the proper quantities? The customer isn't responsible for that—I am because I underestimated. My markup covers me when I make a mistake; otherwise, I lose money.

Unless of course I was smart enough to build in a nice cushion. I don't always use the cushion—but it's there if I need it. When I don't need it, I make a nice profit.

What This Means to You

This is an area where you can save a lot of money. When I buy your materials, I have to build in that cushion to stay profitable. If you buy your materials, I can't mark it up, and you'll save a lot of money.

Before you consider doing this, ask yourself if this step is for you. You can save a lot more money buying your own materials, but if you buy poor-quality materials, or not enough materials, it is now your mistake. You must pay the difference in cost. If you are the kind of person who can't accurately estimate your materials, then let the contractor buy them.

Let's say that you're the kind of person who knows exactly what you want. If so, I suggest buying your own materials.

Where to Get Inexpensive Construction Materials

An incredible way to buy supplies, fixtures—you name it—is eBay. If you want it, eBay has it, for up to 90% off regular store prices.

A friend of mine once spent hours trying to find something eBay didn't sell. I am sorry to tell you that if you want livestock or other living animals, you're out of luck. For the rest of us, if it's legal, you can find it on eBay.

This is a great way to find rare or unusual materials and antiques. If you want a home that is truly unique, eBay will supply you with interesting materials that few people will have.

If you aren't an eBay junkie, a word of warning: Shopping on eBay takes time. If you are in a hurry, you won't save as much money as you could otherwise. Set aside a little time each evening to see what is available. Learn how much things are worth. With a little patience, you will buy good materials at great prices.

Buy Materials from Stores in Urban Renewal Zones

What? You don't know what an urban renewal zone is? I'll tell you.

To put it simply, an urban renewal zone is a place that the government is trying to rehabilitate. These are not the nice, popular parts of your town. These are the places where few businesses have set up shop, for whatever reason.

I know what you're thinking. "Matt, you're sending me into the ghetto." Not necessarily. The neighborhood doesn't have to be dangerous to be a candidate for urban renewal. I know a portion of Brooklyn that used to be strictly industrial before the heavy industry moved away. The city stopped making money.

Gradually, people started moving into the neighborhood. This brought in some money, but not very much because few people moved into the area—only those who were priced out of the nice parts of town. Otherwise, no one wanted to live there. The area was ugly, and far from everything.

When the government labels an area as an urban renewal zone, it gives price breaks to the businesses that agree to move in. Their rent may be very low. This lowers their overhead and the costs the business has to pass on to you. And—more importantly for you—the government allows the business to charge half the sales tax. While other businesses have to collect the full amount of sales tax, these guys can legally collect half.

That means you will spend less money. Just in case you weren't paying attention, this is a good thing.

This is a little secret that I learned by accident. The government is trying to attract businesses into urban renewal zones; therefore, it advertises to businesses, not to private citizens. Do you know which areas in your community are urban renewal zones? No? Here is a handy little website you will want to check out: www.hud/gov/cr. Or you can call 800-998-9999.

This will allow you to locate the urban renewal zones in your area. Depending on various factors, these could also be labeled "Enterprise Zones" or "Empowerment Zones."

You will also want to ask about the sales-tax breaks that your state may allow businesses in these zones to give you. This number varies from state to state.

Buy Loss-Leaders

Stores will often sell particular products or materials at or below their selling cost. Why do they do this? They want to bring you in for the low price, hoping you will also spend money on more profitable items. You can make out like a bandit if you buy the loss-leader and go somewhere else for your other purchases. If you see something else at a reasonable price, comparison shop. Just because the store has a great price on bathtubs doesn't mean they have the lowest-priced tile.

Buy Floor-Models, Damaged, Discontinued, or Overstocked Items

I know, this doesn't sound attractive. And we've all seen stuff in the clearance section that should have been thrown away. But this is not always the case, especially with large pieces of furniture, fixtures, or appliances.

These pieces have to be perfect to sell for full price. That means that a washer with a small dent on the side, for example, must be reduced in price. The same goes for floor models that have been handled by the public. A friend of mine once bought a beautiful wrought-iron chandelier for half-off because it was slightly off balance. He weighted the light side and the problem was solved.

The fact is, you can find goods with "damage" so minor it's silly. Another friend bought a top-of-the-line washer in the "damaged" section two years ago. Do you want to know what was wrong with it? So do we. There are no scratches. And it works perfectly.

Resources: www.overstock.com; www.buy.com.

Buy Refurbished Items

Sometimes people buy products, open the packaging, then change their minds. Or if the product had slight defects, the manufacturer will fix the defects and try to sell it. But now the product can't be called new, so they have to sell it as "refurbished" for a fraction of the retail price.

Resource: www.refurbdepot.com.

Buy Remnants

Remnants are a great value if you only need small pieces of materials. Think of carpet, linoleum, tile, and even wallpaper as good remnant candidates. Home improvement stores sometimes carry other remnants that they will be willing to sell for a deep discount. Ask.

Reuse Materials from Demolished Buildings

Old buildings that are being demolished or remodeled often have materials that can be reused. They can even be of better quality because they were made from materials builders can no longer access. For example, chopping down old-growth trees is a no-no these days. But it didn't used to be. And the wood from old-growth trees is superior because it contains less moisture and is less likely to warp or buckle.

Some materials you might be able to salvage include:

- Crown molding, trim, cabinets, staircase woodwork
- Built-in furniture (i.e., bookshelves)
- Kitchen cabinets and countertops
- Fixtures (lights, sinks, toilets, door knobs, shower doors)
- Fireplaces, fireplace mantels
- Decks, gazebos, bricks, decorative outdoor woodwork
- Hinges, handles, mirrors, and windows

This is not a complete list. The possibilities are endless, and if you go this route, you will have an interesting, unique home built for less than the cookie-cutter house next door.

Have Your Products Delivered as You Need Them

If you have products delivered as you need them, then you won't have to pay for all of the products all at once. You will pay for them as they are installed on your home.

Other advantages to this method include:

- You will have less stuff lying all over your yard. Your spouse or partner will complain less about the mess, and it's safer for any children in the home.

- Rain can't damage your wood if it hasn't been delivered yet.

- People can't steal what isn't there. I know, it's unpleasant thinking about theft, but the sad fact is that if you leave stuff lying around, it's a temptation to all sorts of people.

Be Flexible about Colors and Styles

The color of items can have a big affect on the cost. In general, you will pay less for common colors. Does that sound boring? Well, when was the last time the color of your bathtub kept you up at night? Sometimes we fixate on things that just don't matter. Do you ever really notice other people's trim? If you did, it was probably because someone tried to get creative but ended up doing something strange.

That doesn't mean you should pick what everyone else picks. If you want a custom home on a real-world budget, pick one or two things that are unique and leave the rest alone.

Pick standard colors. Red is expensive just because of the way manufacturers have to make the color. Black toilets are harder to get than white. If you can go with a standard color or standard trim, then you will have more success getting discounts.

Get Thirteen Bids on Materials

Pull out your yellow pages and start calling lumber yards, plumber supply stores, landscaping supply yards—whatever it is you need to buy. Some of these places won't serve the general public, but some of them do. Find out. Ask questions.

Call thirteen places and get a quote on how much they are going to charge you for what you need. Then call Home Depot or Lowe's. Both stores have policies to beat the competition's price by 10%.

Now you may wonder how any other supply yard stays in business if Home Depot and Lowe's are always willing to undersell everyone. Good question. Let's use an illustration.

Have you ever shopped at Wal-Mart? You've probably noticed that their prices are very low. If you went looking for a men's suit, you could probably buy one for $70 to $80.

Now think about the high-end department stores in your area. The minimum price for a suit there is $300 to $500. Why is there such a difference in price?

The quality. I'm not knocking Wal-Mart—if you like to shop there, go for it. All I'm saying is that the price has something to do with the quality. If you want to remodel your house for the cheapest possible price, then a big-box home improvement store will usually deliver that for you.

Personally, I value a high quality product. I like to know that I am using high-end materials that are going to last a long time. I'm willing to pay more money for it. And many of the high-quality products I want can't be found at big-box stores.

That's why you need to know—ahead of time—what kind of project you want to build. What do you value? A low price, or a high-quality product? Choose. And be prepared to be consistent with your decision. This doesn't mean that you need to spend obscene amounts of money to have a high-quality job. I've already told you how to save money on contractors and on materials. If you get thirteen bids on all of your materials, what do you think will happen?

Everyone repeat after me: "I will save a lot of money."

Not only that, but you will have the satisfaction of knowing that you are a savvy consumer. You bought exactly what you wanted for a fair price.

Other Ways to Save Money on Materials

Use materials that are readily available in your area. As a general rule, anything you have to special-order is going to cost you. If you pick materials that are either in stock or not that far away, you will not have to pay for any added transportation or delivery of the materials. Finding out what is readily available in your area will take a little work, but it is worth it when you are looking to get the most for your money.

Return leftover materials. You may have materials you didn't use on the job. Find out *before you buy* if the store allows you to return unused material. Buy from stores that are willing to work with you on this. Just make sure you keep the materials protected from the elements and return them as soon as possible.

Use man-made materials instead of natural materials. Some man-made materials are cheaper than the real thing. Marble is nice, but it is expensive. There are some man-made products that look a lot like marble and cost much less. The same goes for cultured stone and a whole host of products. Ask about possible substitutions.

Use factory-built components instead of on-site labor. Factories use assembly lines and mass-produce their goods. This allows them to charge less for their products than your contractor can build them because he has to build things one at a time with a small crew. Not all ready-made products are of low quality. Do some searching and see if there is anything you like.

Use standard sizes and in-stock materials. Again, if you can buy something in a size that the material supply yard stocks on a regular basis, you can save a lot of money. This tip also works for the shape of your home. Rectangles are cheaper to build. It's best to round up your dimensions to even numbers. Have your home size rounded up or down by increments of two feet to cut down on the amount of wasted materials. Try to keep the depth of your home to thirty-two feet or less. Any more than that and you may need specially designed roof trusses. And that means more money.

Build up instead of out. That way you won't have to build a bigger foundation. Trust me; this will save you money and stress. Besides being cheaper per square foot, your property taxes are less when you build up instead of out.

Work with people or stores trying to break into your market. These people are hungry for your business. They are more likely to give you a better price because they want to build their reputation as quickly as possible. This also works well with contractors who have a lot of experience but may have just moved into the area.

Wrap Up

No one is going to give you a price break just because you are a great person, but if you do your research, you can save money on every facet of your remodel project.

Remember to get your contractor to qualify your project as a capital improvement instead of a repair. An improvement is sales-tax exempt in many states. A repair is not.

Once you do that, consider buying all of the materials for your job, or at least the more expensive pieces. You can find unusual, rare, or just plain inexpensive materials on eBay. If you buy your materials from a yard that is in a designated urban renewal zone, you will only pay half the tax everyone else pays. Just make sure you know enough about what you are buying.

And lastly, don't forget to get thirteen bids for all the different types of materials you need to buy. That way, you will be sure to pay the lowest price possible.

These are all good secrets, yes?

If you stopped reading right here and used all of the secrets I've shared, you would save thousands of dollars. One of the next secrets, should you choose to use it, can save you thousands of dollars *all by itself*. Before we get to that, however, let's discuss ways you can save money on your home expenses *every month*. Are you ready?

How to Save Money on Your Monthly Home Expenses

In this book I am showing you ways to save thousands of dollars on your remodel project. But if you're smart, you won't stop the savings there. Building an energy-efficient house pays. According to the Department of Energy, it is possible to reduce the energy consumption of a new home by 50% without much affecting the cost of building the home (http://www.eere.energy.gov/buildings/info/homes/index.html).

What's that you say? You aren't building a new home? Even if you are only remodeling your kitchen or basement, you can still use energy efficient materials. Use double-paned windows, buy energy-efficient appliances, and insulate your home wherever possible.
These days, with the rising prices of oil and electricity, having an energy-efficient home will also reduce the cost of owning it.

Energy Star®

The Environmental Protection Agency (EPA) has a program called "Energy Star" that rates materials and appliances for energy conservation. Equipment that passes their tests as energy efficient receives the Energy Star® designation. Equipment with this designation can reduce your heating and cooling costs by 15-40%. Some manufacturers go one step further and offer special discounts or lower financing rates when you buy Energy Star®- approved equipment. For more information, go to www.energystar.gov.

Energy Grants and Credits

If saving money on the monthly cost of your home isn't enough, you may also be eligible for grants or tax credits from the government for using or buying energy-efficient products.

California has a program that helps you pay for part of the cost of using energy devices that don't create pollution. Their best-known grant is for using solar power. Check with your state and federal government to see what programs they have for people in your area. Also check with your bank—you may qualify for a lower rate when you finance your home.

Look at the Big Picture

You may need to make a large initial investment to make your house energy efficient. If it costs $6,000 to make your home efficient, you may be tempted to skip it, but that would be a mistake. You need to look at how much that initial investment will save you each month. How long will it take you to recoup your investment? Think of your last heating and cooling bill. How much did you spend? How much would you spend if that bill was cut in half? You may find that your investment has paid for itself within the year, and is now actively saving you money. So don't just assume that energy-efficient products are too expensive. You may be passing up a great way to reduce your monthly expenses.

How $180 Can Save You Thousands

Your greatest savings will come from this next secret. If you want to save thousands of dollars on your home improvement project, start your own business.

Do you think starting your own business will be too complicated? You're wrong. Each state has its own rules, but basically you need to get yourself a business license, a business checking account, and some business cards. The hardest part is trying to come up with a snazzy name. All in all, you'll spend around $180 to make yourself legal.

For specific information for your state, go to your state's official Web site. Find the link for Business. There should be a section on "starting a business" somewhere nearby.

It took me about four hours to start a business. I spent some time searching through my state's Web site looking for what I needed. Then I ordered the forms online and downloaded them. Once I filled out the two copies of the business license forms, I had to take it to a notary to sign. After that, I went to City Hall, gave them $120, and I was in business. No fuss, no muss.

Starting your own business can save you so much money—and open the way for you to make money as well. Opening a business is worth it for how much you will save on your taxes alone. But before we get into that, let's answer a few questions.

Don't You Need to be Licensed to Do Business as a Contractor?

Yes and no.

If you are going to work on your own property, you don't have to be licensed. The property is yours. You can modify it without first becoming a licensed contractor. Now, that doesn't mean you don't have to get building permits. It just means you don't have to take a vocational class before you can build a deck.

You do need a contractor's license if you do work on someone else's property for money. This makes sense when you think about it. You have the right to modify your own property. If you do a bad job, you aren't going to sue yourself over it. But if you are going to build something for someone else's benefit, it has to be up to code.

Think of it this way: If you don't feel well, you have the right to take whatever (legal) substances you like, but you had better have a license and know what you're doing if you prescribe medicine for someone else.

Now, before anyone gets confused, let me make one thing clear: Getting your business license is completely different from getting your contractor's license. Anyone who wants to act like a business must have a business license. A contractor's license is for people who want to work as a contractor on someone else's property.

Everybody clear on that? Good. Now that we've taken care of that, you might be wondering …

Shouldn't I Know What I'm Doing Before I Become a Contractor?

Yes you should. But you really don't have to know that much to get started.

I don't want you to think you'll be able to wire your house after paging through a do-it-yourself book. That will get you killed. Ditto for plumbing and heating. These tasks are too complicated to mess with without proper training.

The position you want to fill is General Contractor. Remember this guy? He's the one who gathers all the necessary permits and hires the subcontractors to do the parts of the job he can't do. This is something you can learn to do fairly quickly.

Do you need to know which permits you need to get from the city? Go to your state's Web site and look it up. If you need to talk to someone, they will list phone numbers on the site. You don't need to have a license to get the permits for your own home.

Once you pay for the permits, your project is on the city's radar. Inspectors will come by to make sure that your project is built to code. If something is wrong, they will point it out to you.

If you don't have a construction background, you can simply hire the people necessary to do the job for you. You can act chiefly as the administrator for the project. This means that you don't need specialized knowledge. Let the inspectors check that your project is being built correctly. That's their job. They're good at it.

You will still want to educate yourself.

How to Educate Yourself

Read. I hope you took my advice and read some of those do-it-yourself books. They will give you an overview of how certain projects are built. Now you want to do some reading to learn construction talk—that specialized terminology that separates the construction professionals from the consumers.

Go to websites like www.nahb.org (The National Association of Home Builders) and start reading their articles and buying their books. If you want to be taken seriously as a contractor, you need to know the lingo.

I recommend *Carpentry and Building Construction* by John Feirer and Gilbert Hutchings. It's an excellent resource. You should also have a trade dictionary so you can look up words you don't know.

Attend home-builder shows. NAHB puts on a good one. You will learn so much from a home-builder show. These events are well worth your money to attend.

You need to know someone in the home-building business to get into an event like this. Unless, of course, you are in your own business like I recommended; then you can easily get into these shows. Either way, it's worth the trouble. I've helped some people attend home-builder shows, and they've thanked me for it. You'll be happy you went.

Go online. There are companies who will help you become your own general contractor. They will give you advice, assistance, and will answer questions. One company that does this is Creative Home Services (www.creativehomeservices.com). They can help you line up financing and materials. They also are a great source of useful information.

Volunteer. A lot of people out there don't have homes. Organizations like Habitat for Humanity build homes for these people. You can volunteer and learn what is involved in building a house. It's on-the-job training, and you're doing some good in the world.

What If I Like Contracting?

You might like working on houses. I know a man who, after acting as the general contractor on his own home, decided to become a builder. He didn't know anything when he started working on his own house. By the time he was done, he still wasn't an expert, but he knew how to do the paperwork necessary to get a house built. The man went and got his builder's license. Now makes good money building houses.

How can you get your license? Each state has different requirements. Visit www.contractor-license.org to find what is required for your state. You can also find this information on your state's official Web site. I like this site because you don't have to dig through a host of other information to find what you want.

In general, here are the things you will have to have:

1. **Some real-world experience.** Acting as the general contractor for your own home counts, as does working for Habitat for Humanity.

2. **A sponsor.** Not all states require this, but it helps to have one. The sponsor basically says that you know what you are doing. A lot of people ask a friend or relative to "sponsor" them. A better way would be to attend a vocational class and have the instructor sponsor you. Finding a vocational school in your area is very easy. Search the Internet under "vocational schools" or "trade schools" and you will find many Web sites dedicated to locating a trade school in your area.

If you want to be taken seriously by contractors,
you need to know the lingo. Educate yourself.

3. **A host of other paperwork.** Check your state's Web site. You may need to be fingerprinted, bring a photo or yourself, and/or present a copy of your business license.

4. **Pass the contractor's test—for those states that require it.** These tests are state specific. Call your state's Department of Consumer Affairs to find the locations for these tests and information on where to purchase study guides. You may be fortunate enough to live in a state that doesn't require a test. Many of these tests are open book. Some sites, like the one listed above, will sell you highlighted, tabbed books for the test. Other companies will coach you before the test so you can pass.

Once you turn in all the required paper work to the city and pass your test, you are in business. You are licensed to do business in the state where you passed your test. This may sound like a lot of work, but it isn't. Like getting a business license, the hardest part is collecting everything together. And that isn't very hard.

Why is starting a business so easy?

The U.S. Government is Pro-Business.

Maybe you know this already. Most people know in a vague way that businesses enjoy more perks than a private citizen. It is hard to see how many perks there are when you work for someone else.

When you open for business, a whole new world is available to you—discounts, tax breaks, and easy-to-get lines of credit. Business is the back bone of this country, and Uncle Sam will reward you for beginning a money-making enterprise.

It used to be that taking a home office deduction brought you special attention from the IRS. No longer. Home businesses are growing at an unprecedented rate in this country. The government has moved with the times, easing the regulations surrounding home businesses.

You can run your business from home and report a loss three years in a row before Uncle Sam will shut you down or demand you make a profit.

Why Does Uncle Sam Care?

Uncle Sam charges this little thing called income tax. He can't get his slice out of your business until you have a profit and report income.

That is why the government takes a dim view toward people who label their hobbies "businesses." They are getting all the advantages that come from running a business without giving Uncle Sam his share of the pie.

In other words, they're cheating.

How to Avoid Getting Labeled as a "Hobby"

Let's be very clear on one point: I am not going to show you how to cheat the government. The secrets I am about to share with you are completely legal. If you are looking for ways to hide income, go somewhere else.

The difference between a hobby and a business, according to the U.S. government, is intent. A hobby is something you do for fun. A business is something you do to make money.

That doesn't mean you can't have fun in your business. We're talking about its *primary purpose*, the point of the whole endeavor. Are you trying to make money? Would you stop the activity if it weren't profitable?

Most businesses eat money for the first few years. Fortunately the government knows this, and you have some time to play. Again, if you can prove that you intend to make money, the government will give you time to make it work. How do you prove that you intend to make money? And why is it important?

The following section has been adapted from the IRS' Website (www.irs.gov) "Frequently Asked Questions" in the Business section.

Hobby expenses are deductible only to the extent of hobby income. In other words, if you make $500 from your hobby, you are only allowed to deduct up to $500 for any expenses your hobby incurs. A business can deduct expenses from your other income sources, so it's important to distinguish hobby expenses from expenses incurred in a for-profit activity. The IRS uses the following nine points to distinguish between hobbies and businesses.

None of these points are more important than the others. If for some reason you attract the notice of the IRS, all of your specific facts and circumstances are going to be looked at as a whole. This list of nine gives you an idea of what they are looking for.

Nine Factors the IRS Uses to Distinguish a Hobby from a Business

Do you carry on the activity in a businesslike manner? The more professionally you conduct your business, the better. Do you have a business checking account? Do you keep good books? Do you have a plan to generate more income?

Does the time and effort you put into the activity indicate you intend to make it profitable? How much time do you spend at your business? The more time you spend, the better. This shows the IRS that you are serious about making money. How many jobs have you completed? How much did the jobs bring in?

Do you depend on the income produced by the activity? Do you use the money you make to pay bills? Would it hurt you financially to lose this income stream, or is it "fun money"?

Are your losses due to circumstances beyond your control (or are normal in the startup phase of your type of business)? Uncle Sam understands business losses—if losing money early on is normal for your type of business. If your business lost money because of other factors, can you document them?

Do you change your methods of operation in an attempt to improve profitability? Have you tried different ways to make your business more profitable? What were they?

Does your team know how to operate a successful business? Have you operated your own business before? If not, have you consulted people who have?

Were you successful in making a profit in similar activities in the past? Have you operated this type of business before? Did it make a profit?

Did the activity make a profit in some years? Has the business been profitable before? How much did it make? Businesses can go through rocky times. Just because you are not making a profit right now doesn't mean you're a hobby.

Can you expect to make a future profit from the appreciation of the assets used in the activity? Some businesses have assets that appreciate. An example: You decide to buy a second house and run your construction business from there. Gradually the neighborhood gentrifies, and your house is worth another $50,000. Your asset just appreciated.

How to Look Like a Business

Here are some things you can do to look more like a business and less like a hobby.

- **Get a business checking account.** You don't want your business and personal money to mingle. Even if you are the only person in your company, you want to separate your business assets from your personal assets.

- **Write a business plan.** At this stage in the game, your business plan can be very simple. State who you are, what you want to do, and how you are going to do it.

- **Have a mission statement.** A mission statement sets out your business values. Your values could be simple. "I want to remodel homes for reasonable prices" can work just fine.

- **Keep your expenses low.** Don't go crazy spending your money. We're here to save money, not go into debt.

- **Consult with more experienced business people.** The Small Business Association or Score (www.score.org) can be very helpful. Consulting the experts shows that you are serious about your endeavor.

- **Take classes/seminars in your field.** It is a lot easier to prove that you will make money in your chosen business if you can show that you know what you are doing.

Wrap Up

In this chapter we've discussed how easy it is to open up your own business. Four hours and $200 will take care of it.

You don't need to have specialized knowledge or a contractor's permit to work on your own home. Your state's Web site and the trusty city inspectors will make sure your remodel project is coming together the way it should.

If you like contracting, I suggest that you spend the time to get a little more experience. You can take vocational classes. Or learn on the job by helping your friends and relatives build their do-it-yourself projects. You don't need a license if you aren't charging them anything for the help.

We also discussed why the IRS doesn't want people claiming hobbies as businesses, and learned what it is they look for in a business.

Now that you know all of this, how do you make it work for you? And why should you bother opening your own business? We'll cover that in the next chapter.

How to (Legally) Deduct Your Remodeling Project from Your Taxes

The Supplies and Tools You Buy Are Tax Deductible

You have to buy the supplies for your remodel anyway—if you open your own business, you can deduct the expense. If you are planning on doing some of the work yourself, do you have to buy any tools? Those expenses can be deducted, too.

Don't stop at tools. All businesses have a required amount of paper work and office work. Printer paper and other office supplies can be tax deductible as well if you use it for your business.

IRS regulation 1.162-1 states that "Business expenses deductible from gross income include the ordinary and necessary expenditures directly connected with or pertaining to the taxpayer's trade or business."

In English "ordinary expenses" means that anything you have to buy in order to do your job counts as a deductible expense.

"Necessary expenses" include items that can benefit your business. You don't need them to do your job, but they help you run your business more profitably. For instance, you don't need a computer to run your contractor business. After all, you could do everything by hand. But a computer could speed up the paperwork part of your venture, and so could count as a necessary expense.

If Your Business Expenses Exceed Your Business' Income, You Can Deduct the Expenses from Your Other Sources of Income

This is where a business is different from a hobby. If you make $500 with your hobby and spent $600, you may only deduct $500 from your income.

Now let's say you work for someone else during the day and have a side business you do on the weekends and evenings. If your business made $500 and you spent $600, you can deduct the $500 from your side business and the other $100 from the gross income from your day job.

You heard me. You can use the losses from your side business to lower the taxable income from your main job. It's completely legal. How's that for a deal?

Actually, working at another job full time works to your advantage another way, too. When you work for yourself, you are solely responsible for paying your income tax, social security, and Medicare. When you work for someone else, they withhold that money from your paycheck automatically, and the company pays for half of your social security and Medicare.

Do you have to buy tools for your remodel project? If you open your own business, those expenses are tax deductible.

When you are in business for yourself, you pay the whole thing. The good news for you folks who aren't giving up your day jobs is that you can pay these extra taxes by having more money withheld from your paycheck. This way, you don't have to mess with sending in your taxes quarterly. Make your main job do it for you.

You Now Have Access to Wholesale Prices

Take advantage of this. Supply yards with fair prices will give you business discounts. If you get a credit line with the yard, sometimes you can get an extra discount.

All together, you can save 5-25% off of what consumers pay. The discount isn't as big as it used to be. The big box stores have forced supply yards to be more competitive in their consumer prices, but you can still save something. Depending on your values, and after you get thirteen bids, you can take that bid to Home Depot or Lowe's for a 10% discount.

You can do the same with plumbing, electrical supply, landscaping, and cabinet supplies. You can save an average of 60% on these items. Sometimes even more. Most kitchen cabinets get a 50-60% markup. When you have your own business, you can avoid the markup.

Open a Business Line of Credit

Open a line of credit at the places where you buy supplies. Then ask if the store will give you a discount for repaying early. Many stores send out bills toward the end of the month, and some of them will give you a 2% discount if you promptly pay the bill by the 10th of the month after receiving the bill.

You can make this work this advantage even more. If you time your purchases for the beginning of the month, then you might have more time to pay for your materials—and still get the early bird discount.

For example, if you buy something May 30th, the purchase will show up on your June bill, and you will have to pay it by June 10th to get the discount. Now, if you wait a few days and buy the same thing on June 3, then you might not get billed until July. If you pay your bill off by July 10th, you get the discount and extra time. How's that for working the system?

Be Aggressive

No one is going to start throwing money your way just because you started your own business. You have to go out and get the discounts. How?

No one can give you a business discount if they don't know you are in business. Don't be shy about passing out those new business cards you bought. Put your company name everywhere. One guy I know put "Builder" on his checks after his name—as in, "James Smith, Builder." Now he gets better discounts on materials than I do at some supply yards.

There is another sure-fire way to find out if a store or supply yard has discounts for businesses. Ask for them.

Let me say that again. Ask for them. Some store discounts aren't publicized. A direct question always works best. "Are there any discounts for businesses that buy things here?" is a good one. Bring along a business card so the store knows you aren't just looking for a freebie.

This is what I call the "discount mentality." A person with a discount mentality is constantly looking for ways to save money. This approach is different from being cheap. A cheap person will only buy something at the lowest possible price regardless of its quality. This kind of person not only shops at discount stores, they think everyone who doesn't is stupid. Quality just doesn't matter.

Tools aren't the only tax deductions you can take.
Anything you use in your business—computers and office
supplies, for example—are also business expenses.

The person with the discount mentality is different. This person will buy something of high quality even though it is more expensive than the cheaper but lower-quality item. He or she simply insists that the price for the high-quality item be as low as it can be without compromising its quality.

This is an important distinction. Cheap people have it easy. They simply search around for the lowest price for the object they want and buy it. If you subscribe to the discount mentality, you can't be lazy. You have to balance value with price and come up with a good compromise. But—to me, anyway—the money you save and the things you get are worth the price.

This "discount mentality" works everywhere. I know another guy who used this to get a better rate on his hotel room. He was traveling for business. As he was about to check in, he asked the front desk person if the rate they quoted him was the best they could do.

The hotel shaved a little off the total hotel bill, and he was upgraded to a better room. All because he asked.

Wrap Up

This chapter discussed some of the advantages you get if you start your own business. Your materials and tools are now tax deductible. Any costs you incur over profits that you make can be used to lower the gross taxable income from another job. And as a business owner, you put yourself in line for discounts from supply yards, office supply stores, and other places that touch on your business. All you have to do to take advantage of them is to have an aggressive, "discount mentality."

All of these advantages sound pretty good, right? But how does this work for you right now, when you want to remodel your home for the lowest possible amount?

The next chapter will put all of these things together in a real-world scenario and reveal how you can leverage all of this knowledge to save a lot of money. Read on.

Putting the
Discount Mentality to Work

Here is a scenario: Let's say Dan Home Owner wants to sell his house in three years and move to where it doesn't snow in the winter. He calls a real estate agent to come look at his house. He wants to know what he would get for it right now.

Dan lives in a small house in an up-and-coming part of Brooklyn, New York. The agent does a walk-through. Afterward she tells him that his house is worth $467,000.

For this neighborhood, that is not a lot of money. The problem? All the other houses in the neighborhood have two bathrooms and Dan's house only has one. Adding a master bathroom will put his house in the mid-$600,000 range with everything else in the neighborhood. The agent also tells him that if he renovates the kitchen—putting in new cabinets, counter tops, and retiling the floor—he can add another $50,00 to $100,000 to his asking price.

Dan thinks that is a great idea. He has read this book, so he knows he should open his own business before he begins the remodel process. He talks to his neighbor, who is also thinking of adding another bathroom. They decide that Dan will be the general contractor for the neighbor's remodel if Dan's remodel turns out well.

Dan's Search for Quality Help

It's summer when Dan decides to start looking for contractors, but it's a seller's market, and no one wants to come and talk to Dan. All the good contractors are busy on big-ticket jobs. Dan decides to wait. He isn't in a hurry. He has three years to get his kitchen and bathroom done.

Meanwhile, Dan remembered that he should be looking for someone to design his remodel before he starts calling contractors. He spends the summer and fall finding, then working, with an architect. Dan's wife has been looking through home magazines and has identified a popular kitchen style. They visit showrooms to see these styles personally. The architect lists exactly what materials they want.

Dan waits until winter when the traditional remodeling season is over; then, Dan begins interviewing subcontractors. Once he has narrowed his list down to three in each category of subcontractor that he needs, he tells each sub that he is going to be the general contractor for the remodel. He tells them that if the person he chooses does a good job at a fair price, he will use him for the job he is going to do next door.

Dan has done his research and knows construction terminology. He sounds like he knows what he is talking about, so the subcontractors take him seriously.

Because Dan can offer multiple jobs to the contractors, he gets an even better price than he would have otherwise. Between the promise of more work and the slow winter season, Dan can afford to hire a high-end cabinet maker and a well-respected tile guy.

Dan Looks for Materials

After calling different lumber yards that sell cabinet-quality wood, Dan decides to have the cabinet maker purchase the materials for the cabinets. The cabinet maker is better connected and ends up getting a better price for the cabinets Dan's wife has fallen in love with.

Dan turns his attention to the tile for the floor. His wife has been shopping on eBay, and has found an excellent deal on ceramic tile. Dan calls around to other tile stores and discovers that the eBay dealer is a good $200 less. The only problem is the shipping. Dan asks the owner if he has any flexibility in his prices for contractors. He mentions that after this job, he has another job that will need more tile. The tile seller agrees to pay for the shipping.

Because he saved so much money on the tile, Dan has enough money to buy marble countertops instead of the laminate he was going to use. His tile guy is skilled enough to put the marble in as well.

Dan then spends his time calling hardware stores and shopping on eBay to find the lowest possible price for new kitchen fixtures. His wife does the same for the trim and kitchen paint.

Dan and his wife get all the necessary permits and paint the kitchen themselves. They leave everything else to the professionals.

Dan Decides He Likes Construction

This whole process has been very interesting to Dan. He isn't willing to give up his job selling widgets for his company, but thinks construction would be a good side job.

While the construction is going on in his home, he takes two week-long classes on basic carpentry and plumbing. The instructors give him letters of recommendation. He then looks online and finds a company who guarantees that he will pass the contractor's licensing test.

By the time he has paid the cabinet and tile guys for their work, Dan is a licensed contractor. Dan's neighbor is impressed with the remodeled kitchen and decides to redo his kitchen, too. Dan and his subcontractors work on the neighbor's house.

It is Dan's first paying project. Dan gets all the necessary permits and walks the inspectors through the remodel when they come around. This takes even less time because Dan knows exactly what he needs. For very little time and effort, Dan gets a percentage of the profits. Dan uses that money to help pay off the loan he took out to pay for his own remodel project.

Becoming a contractor isn't hard. If you have a little first hand knowledge and a sponsor, you're halfway there.

At the end of the year, when Dan is doing his taxes, he takes his remodel expenses and uses them to lower his gross income from his day job. Because he made less money than he spent, he doesn't have to pay any income tax. He also thought ahead and increased his withholding at his other job to take care of his social security and Medicare responsibilities. Even with this added expense, Dan's tax return is bigger than it was last year.

What Did Dan Get for All of This Effort?

With a little planning, Dan has (1) increased the sale price of his home another $200,000, (2) lowered his gross taxable income, (3) saved thousands of dollars on his remodel, and (4) made extra money on the side. Not a bad return on his investment. Not at all.

With some time and a willing attitude, you can have the same results.

Wrap Up

Did you notice how Dan leveraged every advantage he had to save money? Everywhere he went, he asked if there were discounts for contractors. He mentioned that he could supply more work to all involved.

Money is a contractor's number one motivator. If he knows that he is going to get repeat business from you, a contractor will often be willing to lower his quote. Especially during the contractor's slow season. The extra jobs will make it worth his while.

By the way, this only works when you actually have more work to give to your subcontractors. A good contractor can generally tell the difference between an actual promise and an empty one.

Your promises sound like they're worth something when you know what you're talking about. Read magazines and books on remodeling. Go to www.nahb.org and use the resources link to learn more about remodeling in general. When it comes to construction, a little knowledge pays. A lot of knowledge pays even more.

So take the time to learn about contracting. It's an investment that will make your wallet fat.

Have I convinced you to open your own business yet? Good. The next chapter will introduce you to another way to make money from remodeling homes.

How to Maximize the Amount of Money You Make as a Real Estate Flipper

So far we've talked about ways to increase the value of your own home and how to make some money on the side working on other people's homes. Now we'll discuss another way to make money on houses in need of repair.

Perhaps you know what a real estate flipper is. It seems to be the hot new thing to do with a home. Simply put, a real estate "flipper" buys a house, fixes it up, then sells it at a profit.

Sounds easy, right? We all know that renovating is one of the quickest and easiest ways to add value to a home. That was one of the reasons you want to remodel your own home, right? If you have to sell it, you want to make money.

There are people who have taken this concept and used it to make millions of dollars. A certain guy you might know—his name is Donald Trump—made his money in real estate. Now, you aren't Donald Trump, but with a little sweat and ingenuity, you can get a return of $2-$3 on every $1 that you spend. In fact, the majority of really wealthy people have made their money by investing in real estate.

The Catch

You knew there had to be one, right? The catch is that you can't buy any old house, fix it up, and sell it at a profit. There is a method to finding the right place, putting the right amount of money into it, and selling it at the right time. That subject takes whole books of its own. Since this book is about saving money on remodeling, this isn't the book to talk about it. If you want to learn more, do an Internet search or go to your local library and book stores.

What we can talk about is one general point—basically, if you are
going to get into this business, you need to make renovations at the
lowest cost possible without compromising quality. If you have to
do more than cosmetic surgery on a house, then you shouldn't have
bought it.

How to Make an Investment Property Look Great for Less

Most buyers are looking for a lot of space. They want to get as much
space for their money as possible. Even if they can't afford a big
house, they want to feel like they are living in something big. If you
can make your property look and feel big, you will have more buyers
at a higher price. How do you do this?

Use light colors. Dark paint may be ultramodern, but it calls
a lot of attention to the walls, making a place feel smaller.
Lighter colors give a sense of space. Even if you chose to
paint the walls darker for some reason (say, in a family room
where you want a "cozy" feeling) then keep the ceiling light.
You don't want buyers feeling claustrophobic. Pick fabrics in
lighter colors, too. All of your blinds and trim should be light
colored as well.

Raise ceilings. Another way to create the feeling of space
(especially if you don't actually have any) is to raise flat
ceilings. Turn them into cathedral ceilings by opening up attic
areas. It is an inexpensive thing to do, and has a lot of wow
appeal.

Do the work yourself. All of the money-saving strategies in
this book work for investment properties. Doing the work
yourself—either as the general contractor if it is a big job or as
the handyman if the job is small—will save you extra cash.

Give the house a professional cleaning. I mentioned this in an earlier chapter, but it is even more important when you are trying to make money on an investment property.

Replace all of the fixtures and trim, then paint. This will make the home look newer than it might actually be.

Wrap Up

Flipping real estate is a complex subject that can net you a lot of money. Once you have researched the subject, use these tips to minimize how much money you put out and maximize your selling price.

Make My Money Saving Secrets Work for You

There you have it—dozens of contractor secrets that will allow you to shave thousands of dollars off your home remodeling project.

Flip to the table of contents. Thumb through the book and review these secrets again. There is a lot of information, but the principles are easy to follow. So easy, I know someone out there is thinking, "That's it? This is the information that is going to save me money? But it's so easy."

I told you that saving money on your home remodel isn't rocket science. Anyone can take advantage of these closely held contractor's secrets and make them work. Anyone. Well, almost anyone.

The fact is, this book contains real strategies to save money—and real strategies require real work. This isn't a "make money while you sleep" miracle cure.

It takes time to gather thirteen bids for each contractor. Time to search out all of the supplies you need for your job. Checking contractor references and educating yourself takes effort. And sitting back before you do any of these things and planning your project step by careful step takes patience.

If you aren't willing to be patient, and put in the time and effort, then my secrets won't work for you. If you are the type of person who doesn't follow through, then reading this book is a waste of time.

I have given you information the average home owner will never know, but in the end, you have to decide to do something about it. If you are a doer, then this book is for you. You know the secrets. Now use them.

Did you use my secrets? Did you come up with new ones? If

so, contact me at info@homecontractorsecrets.com or at www.
homecontractorsecrets.com. Let me know how much you saved. I
may just use your story in the next edition of this book. If I do, I'll
send you a free copy of the book, and a $50 reward.

Once You Save Money, Give Back

I know that you will save a lot of money if you use the secrets in this
book. Once you do, help someone else. This is the single action that
separates the well-off from the greedy.

There are so many groups that can use your help. You can join the
Rotary Club, Kiwanis, the Red Cross, the United Way, and Habitat for
Humanity. They're always looking for volunteers.

You can find many organizations like this in your phone book,
through your local church, or at the hospitals near your home.
Would you rather work with animals? Your local animal shelter
needs people, too.

I'm not preaching. If you have disposable income because I helped
you, then pass the love along. That includes charitable giving.
Look at the world's wealthy. They know that giving back leads to
happiness—and wealth of a whole different kind. It's the principle of
sowing and reaping.

God bless you and happy remodeling.
Matt Miglin

Appendix A

Labor Payment Methods

Once you settle on one particular contractor, you are going to have to hammer out a contract. As I said in an earlier chapter, you need to describe IN DETAIL how you are going to pay the contractor. This is your primary responsibility as the home owner.

This appendix will give you an overview of the different payment methods you can use and how that affects the cost of your project.

Job Cost

A job-cost bid is a flat fee for labor and materials. This is great when you want to know exactly how much the project is going to cost. There can be cost overruns, but not as many if you plan everything on paper ahead of time.

This type of bid works best when building new homes, on "remodels" where everything but one wall is knocked down, or where the project is clearly defined.

Time and Materials

A time and materials bid means the contractor will bill you for materials and charge you for labor based on how much time it takes him to complete the project. This means you don't know how much your project is going to cost until it's over.

On one hand, you may get a better quality job because the contractor isn't in a hurry. On the other hand, the contractor isn't in a hurry. He makes money by taking longer even if he isn't doing anything, so the costs can get out of hand.

Why would you use this type of bid? This is sometimes the only way to get a bid on a project that is hard to determine before you start working on it. Let's say you have an old house that you are trying to renovate/restore. The contractor won't always know what he's in for until he opens up the house and takes a look inside. This is especially true if he has to open walls or move them around. If you try to make him give you a flat job-cost bid, he (if he's smart) will give a worst-case scenario to cover unexpected problems. Your best bet will be to put in the contract that the project can't go over a certain dollar amount.

Time and materials might be a potential money hole when you are dealing with contractors, but if you are working with an architect, it can be the smart way to go— especially if you aren't using all of the architect's services. Let's say you buy stock plans and want to make a few changes on them—or you simply want a more detailed list of materials. You can save money by paying for a few hours of time instead of a flat fee.

Percentage Basis

A percentage basis means you are paying someone based on a percentage of the total cost. If you chose to use the full services of an architect, this is generally how they determine their price. The average is around 5-12%. If your remodel costs $200,000, for instance, then the architect's fee will be around $24,000.

This is a good way to hire someone when you are looking for performance-based results. It's also good for projects that are complex, and if the contractor can't determine how many hours it will take. Unfortunately, it can be the most expensive way to hire labor.

Hourly Fee

This works about the way you think. You are paying someone for the time they spend on your project. This can be a good way to go if the project is small, clearly defined, and can be finished in a short amount of time. Don't use it for long, complicated projects.

Fixed Fee

This method gives the contractor or other expert a set fee for their labor. These are different from job-cost bids because they only deal with labor and not materials. This method can work for projects where you are providing all of the materials, and all the contractor has to do is show up. On the other hand, if the project hits some unexpected complications (or you change your mind about something), then the fee can change, and may have to be renegotiated.

Price by Square Footage

Sometimes a draftsperson, designer, or architect will charge you a price based on the square footage of your project. The bigger the house, the more expensive the cost. On the other hand, this can be much cheaper than paying a percentage of your project's total cost.

Payment Schedule

If you follow the advice I give in this book, you are going to give the contractor small payments all the way through the construction project. The contractor will often ask you for a deposit up front. This is reasonable, and shows your intent to pay on time from the beginning.

Beware of any contractor who wants an excessive deposit up front. Most states limit how much the contractor can receive up front to no more than 30%. If you are going to pay the contractor often, you can usually get him to agree to a smaller up-front deposit.

Do not make the final payment until after the contractor, subcontractors, and suppliers sign a lien waiver. A lien waiver basically states that the contractor has paid for the work and materials used to build the home. It also means you will not be held liable if the contractor doesn't actually pay those debts.

If you have a construction loan, make sure that you have to sign off on all payments before the contractor gets the money. Sometimes a contractor asks to be paid for work he hasn't yet done. I shouldn't have to tell you that that is a bad idea. Pay only for work that has been done. The ONLY exception to this is the initial down payment.

Wrap Up

This, in a nutshell, covers the more common payment methods you will run into. Pick the payment method that best suits your project.

Appendix B

More Details on Contracts

I'll say it again. Have a contract. A contract is your first defense against a shady contractor. Doing business without a contract is like jumping out of an airplane without a parachute and hoping for the best. Eventually you are going to crash and burn.

A Handshake Isn't Enough

Not for a home owner who wants to save money on a remodel. You might think the contractor knows what you want, but how do you know for sure? How do you prove that you told the contractor that he had to repaint all the trim? Answer: You don't. Not without something in writing.

The Danger of Verbal Contracts

Everyone's memory is faulty. And some contractors will take advantage. Verbal contracts are binding in some states. You may think that the contractor is promising to do something. He could think you're just talking. Don't get burned.

Write It Down

Write everything down. This keeps you and the client on the same page. Here are some of the things I like to include in my contracts. This doesn't cover everything. For more specific help, consult an attorney.

Describe every phase of the project in detail. You want to be crystal clear on what the job includes.

Change orders. Specify that all change orders must be written down and signed off by both parties to be valid. Verbal promises do not count. This sounds like a hassle, but it will prevent arguments between you and the contractor. And ensure any extra stuff gets completed.

Termination fees. Describe under what circumstances the contract can be terminated.

A start and end date. This sounds basic, but if you think the job starts next month and the client thinks it starts next week, you have a problem.

Contractor/Client responsibilities. Specify who is responsible for the quality of the materials that go into your remodel. If completing your job on time hinges on something the contractor has to do, write this down. Spell out each person's responsibilities. IN DETAIL.

Provisions. A provision is a condition attached to an agreement. The contractor may say that his work is only guaranteed for two years, for instance. It is more important to state what the contractor's bid **does not** cover than what it does cover. In my contracts, I clarify anything that could be considered part of the job. Not all contractors are savvy enough to do this. You might think that will give you an advantage, but all it really does is create stress and bad feelings. Spell everything out.

Signature. Another basic requirement. Don't give the contractor any money until you get his signature. It takes ten seconds to sign something. And with the handy-dandy fax machine, you can have it in the contractor's hand and back again in ten minutes. Be suspicious of anyone who is slow signing a contract. This is your key to recovering any damages from the contractor if he doesn't do a quality job.

Have the contractor develop a job completion punch list. This is separate from a contract, but just as important. He should do this before his last day on the job. Call this the final walk-through punch list. On this punch list, the contractor should list, in writing, everything that must be done. Both parties should sign it. That way you will know for sure that nothing was missed.

Wrap Up

People who don't want to be taken to the cleaners use contracts. If the relationship between you and the client turns sour, this is your first proof that you operated in good faith. Protect yourself. Protect your business. Use a contract.

Appendix C

Appliance Resources

On some websites you can purchase large and small kitchen appliances for 35% less than if you buy them at your local appliance store. Here's something you might not know: When you order from another state, you don't have to pay sales tax. Often the shipping cost can be less than the sales tax.

LARGE APPLIANCES:

http://www.homeappliances.com
This site has a price and features comparison for dealers in your area.

http://www.appliance.com
This buyer's guide gives a lot of product information and contact information for purchasing.

http://www.allabouthome.com
This site gives repair cost estimates.

http://www.4appliances.com
This site lists appliances and distributors to help you locate the lowest prices.

SMALL APPLIANCES:

http://www.appliances.com
This site sells small appliances at a significant savings.

http://www.vkitchen.com
This site sells discounted small appliances and cookware at up to
40% off.

http://www.lowestdollar.com
This site offers low prices on appliances.

TOOLS:

These sites have new and used hand and power tools at a discount.
Some also have supplies you might need for your particular project.

http://www.re-tool.com
They claim up to 50% off for used tools.

http://www.toolsupermarket.com
http://www.thetoolboxusa.com
http://www.workshoptools.com
http://www.forcemachinery.com

Appendix D

Business Resources

This is not a complete list of what's out there. Do some research of your own to find the lowest possible prices on products that fit your tastes. I am not endorsing any of the sites listed below. This selection is meant as a little boost to your own research, nothing more. This list is in addition to the resources listed within the book.

Business Cards

www.vistaprint.com
Choose your own design, set up the card the way you like it. If you choose your business card design from a certain set, they will be absolutely free. All you have to pay for is the shipping. I spent $10 for fifty cards.

www.overnightprints.com
Another inexpensive place to get business cards. Vista print can be cheaper if you stick with their free designs, but your choices are fewer. This site will give you fifty cards for $25, and the designs are more sophisticated.

Appendix E

Business Basics

Home Contractor's Secrets—Revealed is all about how to save thousands of dollars on your remodeling project. As you know, one great way to save a lot of money is to go into business for yourself. I'm not going to get into details on how to start your own business. That subject needs a book all its own, and there are many good ones out there. Or you can order my book *The Six-Figure Contractor* and learn lots of profit-making business principles to make good money in contracting. Go to www.sixfigurecontractor.com.

What I will give you is a short overview of the different business structures you might want to look into if you do decide to go into business.

What Will I Name My Business?

I suggest using something other than your name. Smith Construction might have a nice ring to it, but it sounds like some guy working out of his basement. There is nothing wrong with working from your basement. But if you decide to go out looking for work, you don't want the client to think he can pay you less because you are a small-time operator.

But don't get too cute. And make sure people can tell what you do from your business name. NYC Home Remodelers might not win you any awards for cleverness, but everyone who sees your business cards will know what you do. And where you do it.

What Legal Structure Will I Use?

This gets a little more complicated. The short answer is that most contractors begin as a **Sole Proprietor**. Once they start making money, they form a **Corporation**. The following is a list of the different types of business structures you could use, and their pros and cons.

The form your business takes depends on a lot of factors—and your choice has long-term implications on everything from how you pay your taxes, to your liability if someone decides to sue you.

Consult with an accountant and an attorney to help you select the form of ownership that is right for you. In making a choice, you will want to take into account the following factors:

- Your vision regarding the size and nature of your business
- The level of control you wish to have
- The level of "structure" you are willing to deal with
- The business's vulnerability to lawsuits
- Tax implications of the different ownership structures
- Expected profit (or loss) of the business
- Whether or not you need to re-invest earnings into the business
- Your need for access to cash out of the business for yourself

Sole Proprietorships

The vast majority of small businesses start out as sole proprietorships. These firms are owned by one person, usually the individual who has day-to-day responsibility for running the business. Sole proprietors own all the assets of the business and the profits generated by it. They also assume complete responsibility for any of its liabilities or debts. In the eyes of the law and the public, you are one in the same with the business.

Advantages of a Sole Proprietorship

- Easiest and least expensive form of ownership to organize.

- Sole proprietors are in complete control, and within the parameters of the law, may make decisions as they see fit.

- Sole proprietors receive all income generated by the business to keep or reinvest.

- Profits from the business flow directly to the owner's personal tax return.

- The business is easy to dissolve, if desired.

Disadvantages of a Sole Proprietorship

- Sole proprietors have unlimited liability and are legally responsible for all debts against the business. Their business *and* personal assets are at risk.

- May be at a disadvantage in raising funds and are often limited to using funds from personal savings or consumer loans.

- May have a hard time attracting high-caliber employees, or those that are motivated by the opportunity to own a part of the business.

- Some employee benefits, such as, owner's medical insurance premiums are not **directly** deductible from business income (only partially deductible as an adjustment to income).

Federal Tax Forms for Sole Proprietorship

(This is only a partial list and some information may not apply to you.)

- Form 1040: Individual Income Tax Return

- Schedule C: Profit or Loss from Business (or Schedule C-EZ)

- Schedule SE: Self-Employment Tax

- Form 1040-ES: Estimated Tax for Individuals

- Form 4562: Depreciation and Amortization

- Form 8829: Expenses for Business Use of your Home

Partnerships

In a partnership, two or more people share ownership of a single business. Like proprietorships, the law does not distinguish between the business and its owners.

The partners should have a legal agreement that sets forth how decisions will be made, how profits will be shared, how disputes will be resolved, how future partners will be admitted to the partnership, how partners can be bought out, or what steps will be taken to dissolve the partnership when needed.

Yes, it's hard to think about a "break-up" when the business is just getting started, but many partnerships split up at crisis times, and unless there is a defined process, there will be even greater problems. They also must decide up front how much time and capital, etc. that each will contribute.

Advantages of a Partnership

- Partnerships are relatively easy to establish; however, time should be invested in developing the partnership agreement.

- With more than one owner, the ability to raise funds may be increased.

- The profits from the business flow directly through to the partners' personal tax returns.

- Prospective employees may be attracted to the business if given the incentive to become a partner.

- The business usually will benefit from partners who have complementary skills.

Disadvantages of a Partnership

- Partners are jointly and individually liable for the actions of the other partners.

- Profits must be shared with others.

- Since decisions are shared, disagreements can occur.

- Some employee benefits are not deductible from business income on tax returns.

- The partnership may have a limited life; it may end upon the withdrawal or death of a partner.

Types of Partnerships that Should be Considered:

General Partnership

- Partners divide responsibility for management and liability, as well as the shares of profit or loss according to their internal agreement. Equal shares are assumed unless there is a written agreement that states differently.

Limited Partnership and Partnership with Limited Liability

- "Limited" means that most of the partners have limited liability (to the extent of their investment) as well as limited input regarding management decisions, which generally encourages investors for short term projects or for investing in capital assets. This form of ownership is not often used for operating retail or service businesses. Forming a limited partnership is more complex and formal than that of a general partnership.

Joint Venture

- Acts like a general partnership, but is clearly for a limited period of time or a single project. If the partners in a joint venture repeat the activity, they will be recognized as an ongoing partnership and will have to file as such, and distribute accumulated partnership assets upon dissolution of the entity.

Federal Tax Forms for Partnerships

(This is only a partial list and some may not apply.)

- Form 1065: Partnership Return of Income

- Form 1065 K-1: Partner's Share of Income, Credit, Deductions

- Form 4562: Depreciation

- Form 1040: Individual Income Tax Return

- Schedule E: Supplemental Income and Loss

- Schedule SE: Self-Employment Tax

- Form 1040-ES: Estimated Tax for Individuals

Corporations

A corporation, chartered by the state in which it is headquartered, is considered by law to be a unique entity, separate and apart from those who own it. A corporation can be taxed, it can be sued, and it can enter into contractual agreements. The owners of a corporation are its shareholders. The shareholders elect a board of directors to oversee the major policies and decisions. The corporation has a life of its own and does not dissolve when ownership changes.

Advantages of a Corporation

- Shareholders have limited liability for the corporation's debts or judgments against the corporations.

- Generally, shareholders can only be held accountable for their investment in stock of the company. (Note, however, that officers can be held personally liable for their actions, such as, the failure to withhold and pay employment taxes.)

- Corporations can raise additional funds through the sale of stock.

- A corporation may deduct the cost of benefits it provides to officers and employees.

- Corporations can elect S corporation status if certain requirements are met. This election enables the company to be taxed similar to a partnership.

Disadvantages of a Corporation

- The process of incorporation requires more time and money than other forms of organization.

- Corporations are monitored by federal, state, and some local agencies and, as a result, may have more paperwork to comply with regulations.

- Incorporating may result in higher overall taxes. Dividends paid to shareholders are not deductible from business income; thus, this income can be taxed twice.

Federal Tax Forms for Regular or "C" Corporations
(This is only a partial list and some may not apply.)

- Form 1120 or 1120-A: Corporation Income Tax Return

- Form 1120-W: Estimated Tax for Corporation

- Form 8109-B: Deposit Coupon

- Form 4625: Depreciation

- Other forms as needed for capital gains, sale of assets, alternative minimum tax, etc.

Subchapter S Corporations

A tax election only, this election enables the shareholder to treat the earnings and profits as distributions, and have them pass through directly to their personal tax return.

The catch here is that the shareholder, if working for the company, and if there is a profit, must pay him or herself wages, and it must meet standards of "reasonable compensation." This can vary by geographical region as well as occupation, but the basic rule is to pay yourself what you would have to pay someone to do your job, as long as there is enough profit.

If you do not do this, the IRS can reclassify all of the earnings and profit as wages, and you will be liable for all of the payroll taxes on the total amount.

Federal Tax Forms for Subchapter S Corporations

(This only a partial list and some may not apply.)

- Form 1120S: Income Tax Return for S Corporation

- 1120S K-1: Shareholder's Share of Income, Credit, Deductions

- Form 4625: Depreciation

- Form 1040: Individual Income Tax Return

- Schedule E: Supplemental Income and Loss

- Schedule SE: Self-Employment Tax

- Form 1040-ES: Estimated Tax for Individuals

- Other forms as needed for capital gains, sale of assets, alternative minimum tax, etc.

Limited Liability Company (LLC)

The LLC is a relatively new type of hybrid business structure that is now permissible in most states. It is designed to provide the limited liability features of a corporation and the tax efficiencies and operational flexibility of a partnership. Formation is more complex and formal than that of a general partnership.

The owners are members, and the duration of the LLC is usually determined when the organization papers are filed. The time limit can be continued if desired by a vote of the members at the time of expiration. LLCs must not have more than two of the four characteristics that define corporations: (1) limited liability to the extent of assets, (2) continuity of life, (3) centralization of management, and (4) free transferability of ownership interests.

Federal Tax Forms for LLC

Taxed as partnership in most cases, corporation forms must be used if there are more than two of the four corporate characteristics, as described above.

Wrap Up

Deciding the form of ownership that best suits your business venture should be given careful consideration. Use your key advisors to assist you in the process.

I am not a lawyer. I can't tell you which business structure will best cover all of your personal needs. But as I said before, most people start out as a Sole Proprietorship and then change into a Corporation of some sort.

If you choose to do the same, I suggest you do some reading on LLCs. I have found an LLC to be a very convenient business model for my contractor's business. You might find the same.

For more information on other ways to save money, check out our Web site

www.homecontractorsecrets.com

You will find:

- *Home Contractor Secrets—Revealed* on audio CD

- Home-improvement consulting

- Money-saving e-tips

- How to cut your electrical bills in half

- How to reduce your heating and cooling costs by 30%

- Informative home-improvement seminars

- And other helpful, money-saving information

To order now, go to:
http://www.homecontractorsecrets.com
or call
1-866-316-3700

About the Author

Matthew D. Miglin is a forward-thinking professional speaker, author, consultant, and business adviser. He offers exceptional business wisdom gained through twenty-two years of business development, training, and management experience, directing operations for several successful entrepreneurial endeavors.

Matthew is the CEO of Covenant Life Enterprises—a training, consulting, and educational resources company that specializes in helping individuals, families, and businesses to prosper, save money, and live more fulfilling lives. He has presented well-received and profit-enhancing business seminars to individuals, businesses, schools, and government agencies.

A third-generation builder with twenty-plus years of experience, Matthew has excelled as president of a multi-million dollar building and remodeling firm. He has a proven track record of increasing profits, production, and doubling sales three consecutive years. Matthew's professional experience also includes training and licensing in the financial services industry. This has enabled him to develop a strong background in real estate investing and money saving strategies.

He developed much of his leadership skills under challenging circumstances while serving in the United States Marine Corps and during Operation Desert Storm.

Matthew's broad cross-section of experience also includes:

- Certified Seminar Leader with the American Seminar Leaders Association
- Adjunct Professor of Business and Marketing at local colleges
- Featured speaker at the NAHB – Annual Home Builders Show
- Business Development Program Instructor for Rutgers University & FFCDC

- Co-leader on two Habitat for Humanity building projects in Georgia & Florida
- 1994/1995 recipient of Who's Who of Business Leaders
- 1996 recipient of International Who's Who of Professionals
- 2001 recipient of International Who's Who of Entrepreneurs
- Best of Show, 1st Place and Outstanding Award winning furniture builder

Matthew received his Doctorate in Business Development from Cambridge International University and Bachelors and Masters degrees in Business Administration from American State University.